D0803164

Continuing the Good Life

HELEN AND SCOTT NEARING

Continuing the Good Life

HALF A CENTURY OF HOMESTEADING

Schocken Books • New York

First published by Schocken Books 1979

10 9 8 7 6 5 4 3 2 1 79 80 81
First Edition

Library of Congress Cataloging in Publication Data

Nearing, Helen.
Continuing the good life.

Includes index.
1. Farm life—Maine—Harborside. 2. Nearing, Helen. 3. Nearing, Scott, 1883- I. Nearing, Scott, 1883- joint author. II. Title.
S521.5.M2N42 974.1'45'04 78-21151

Manufactured in the United States of America

CONTENTS

Continuing the Good Life

"I have written no more than I have seene, nor added a benefit which I have not knowne liberally bestowed upon the industrious; of which, if you will be a partaker, follow their imitation, and to good labours add a good life."

Gervase Markham, Farewell to Husbandry, *1620*

"Let not the Farmer expect Novelties in every Page; let him not wonder if he finds his every-day Practice plainly and concisely related. My Book is not designed to amuse but to instruct, being filled with Truth, not fancies. And Truth is so plain and obvious that in great Measure 'tis known to all in every Age."

John Laurence, A New System of Agriculture, *1726*

"The author offers no special theory, clothed in visions of fancy; his are the proceedings of a man, who, used to move in a respectable sphere, felt the reverses brought about by political causes, and who, as a true citizen of the world, sought the reinstatement of his former circumstances by seeking a place where his diminished means, his personal labour, and the resources of his mind could be actively employed."

Joseph Pickering, Inquiries of an Emigrant, *1832*

"Since I have given my attention to the cultivation of the soil, I find I have no competition to fear, have nothing to apprehend from the success of my neighbour, and owe no thanks for the purchase of my commodities. Possessing on my land all the necessaries of life, I am under no anxiety regarding my daily subsistence."

John Sillett, A New Practical System of Fork and Spade Husbandry, *1850*

"The case of one wholly inexperienced has been assumed in the present instance, and the information necessary to enable him to meet the different emergencies of the farm carefully yet succinctly given in the following pages, so as to form what it is hoped will be found a useful manual of husbandry on a small scale."

Martin Doyle, Small Farms, *1859*

"If he be by nature indolent, and in temper desponding, easily daunted by difficulties and of a weak frame of body, such a life would not suit him. If his wife be a weakly woman, destitute of mental energy, unable to bear up under the trials of life, she is not fit for a life of hardship—it will be useless cruelty to expose her to it."

C. P. Traill, The Canadian Settler's Guide, *1860*

INTRODUCTION

HOMESTEADING AS A PRODUCTIVE AVOCATION

WE HAVE been experimenting for almost half a century with living a simple life: maintaining health and vigor with a minimum of effort and money and a maximum of satisfaction.

Living the Good Life, a book we originally published in 1954, was based on nineteen years of homesteading in Jamaica, Vermont, seven miles up a dirt road to Pikes Falls. Since 1952 we have been homesteading on another farm, along another dirt road, in Harborside, Maine, on Cape Rosier. This is a report on our Maine experiment and experiences.

Both in Jamaica and in Harborside we took over exhausted, derelict farms. During our time in Vermont the Pikes Falls back road was poorly graveled. At the start and for many years the Harborside-Cape Rosier road was of gravel and came to a dead end, the last of all town roads to be plowed in the snowy spring. Both farms were isolated and remote from towns.

Homesteading in Vermont and in Maine taught us many things about life in general and homesteading in particular. Both were part and parcel of one experience: building and maintaining a solvent family economy amid the wreckage and drift of a society that was disintegrating in accordance with the laws of its own self-destructive being.

In Harborside, as in Jamaica, we turned depleted farmland into good soil, enriching it to a point that made raising our own food possible and productive. Both farms eventually provided us with the food, fuel and shelter which are the minimum requirements of life in New England. At the outset in Vermont, the going was a bit rough because we were novices. By the time we reached Maine, we were sufficiently experienced and equipped to make the transition and the new start with a good deal of background.

In our off-seasons (the deep freeze and its approaches) in both homesteads, we busied ourselves by making essential improvements in our housing. Year by year we became better established, through our own efforts and those of friends and chance itinerant helpers.

At the same time we continued to take an active interest in our professions (Helen, in music; Scott, in social science). Both of us have traveled widely. Both have done a considerable amount of reading and study, and taught when and where opportunity offered. Our interests were and are wide and will continue so in the future. An old adage says that change of occupation is as good as a rest. We would say better than a rest, because the changes in our occupations have provided relaxation without any boredom.

One of our sophisticated city visitors asked us: "What do you do with your spare time?" "We have no spare time; we keep busy," was the answer. "As a matter of fact, the days are so short that we run out of time constantly." "But what do you do for pleasure?" our visitor persisted. "Anything and everything we do yields satisfaction. If we didn't enjoy it, we would do something else, or approach our jobs in a way that made more sense."

"Examine any one of our days," we continued to our inquirer, "or any one of the major activities in which we are engaged: food production and storage; the cutting of our wood for fuel;

gardening; building houses; forestry; research; teaching; music making; speaking; writing articles and books; traveling. Each one of these has its own particular advantages and opportunities. When it reaches a climax or leads to a conclusion, we say: That job is done to the best of our ability; now let's see what the next item on the program is and get on with the day's or week's or season's work."

Our lives are not loaded down with sterile repetition or barren routine. Each new project and each new day is a fresh challenge and an exploratory experience, unless we make some stupid blunder and are compelled to rip out the faulty construction and improve on it. If you identify the mistake and find out how and why it was made, you have the satisfaction of doing the job to the best of your ability and avoiding a like mistake in the future.

Does this sound mechanistic and self-satisfied? We don't mean it that way. If you tackle a job that is far beyond your experience and your energy, it may get you down and keep you down. But if you bite off no more than you can chew, and masticate it thoroughly, your chances of success are good.

There is a tendency nowadays to elbow a way through the mazes of a complicated life. Wisely and slowly is good advice. If you are running a relay race, it is not decided in the first few laps. Take your time. Ration your energies. Plan your operation carefully. Take one step at a time. Then prepare carefully for the next step. It pays in the long run.

Is this a repudiation of the Big Leap principle? Far from it. A single step is better than none at all. In a revolutionary situation a big leap may be the obvious and expedient answer for an individual or an entire group. A big leap at the wrong time, or in the wrong direction or with insufficient preparation or insufficient means, may set an individual or a movement back for a generation. With widely scattered forces and a minimum of experience, a step-by-step policy may well be the right tactic.

Actually, what we have been doing during these decades of experiment and construction is to meet a series of challenges, each in its own time and at its own level. Each challenge had its own peculiar character. Some of them dealt with our basic assumptions; others involved minor practical details. Each was interesting in its own way, and each when solved or resolved yielded its own measure of satisfaction. Each challenge met and mastered adds zest to the present moment. Each one opens up interesting prospects for the immediate and the more distant future.

Even at our advanced ages (both of us are far past the point of customary retirement), we have no desire to withdraw from life. On the contrary, we are eager and anxious to live. The past has been replete with zestful experiences. The present opens up interesting prospects for the future. Life for us has been rewarding, even in its details and minor incidents. We have every reason to suppose that it will continue to yield greater satisfactions as we grow wiser and more competent to deal with what the future has in store.

photo: Richard Garrett

"The breaking waves dashed high on a stern and rock-bound coast,
And the woods, against a stormy sky, their giant branches tossed....
The ocean-eagle soared from his nest by the white wave's foam,
And the rocking pines of the forest roared; this was their welcome home!"

Felicia Hemans, The Landing of the Pilgrim Fathers, *1620*

"Bustle, bustle, clear the way, He moves, we move, they move today;
Pulling, hauling, father's calling, Mother's bawling, children squalling,
Coaxing, teasing, whimpering, prattling, Pots and pans and kettles rattling,
Tumbling bedsteads, flying bedspreads, Broken chairs, and hollow wares
Strew the street—'tis moving day.

Bustle, bustle, stir about, Some moving in—some moving out;
Some move by team, some move by hand, An annual collithumpian band.
Landlords dunning, tenants shunning; Laughing, crying, dancing, sighing—
Spiders dying, feathers flying, Shaking bed rugs, killing bed bugs,
Scampering rats, mewing cats, Whining dogs, grunting hogs,
What's the matter? Moving day!"

Peter Parley's Almanac *for 1836*

"Moving yet and never stopping, Pioneers! O Pioneers!"

Walt Whitman, Leaves of Grass, *1865*

"O fortunate, O happy day,
When a new household takes its birth
And rolls on its harmonious way among the myriad homes of earth."

Henry Wadsworth Longfellow

CHAPTER 1

WE MOVE, BAG AND BAGGAGE, TO MAINE

WE HAD built up our good life in Vermont, improving the soil, clearing out and enlarging the sugar orchard, replacing wooden shacks with concrete and stone buildings, reconstructing the roads and generally converting a sickly, bankrupt farm into a vigorous, healthy enterprise that was paying its own way and more. Several of the friends and visitors who stopped in to see us said quite early: "You have something going here which is very special. When you want to sell the place, let us know before you put it on the market."

We thanked our prospective buyers for the compliment they were paying us, but assured them that we had put much thought, time and energy into the construction and furnishing of our Forest Farm homestead, and selling out was the last item on our list of priorities.

We were overlooking one of the most implacable taskmasters that dog the steps of mankind: the effect of change. We were juvenile and soft-headed enough to believe that we were settled in Pikes Falls, Vermont, forever. We were wrong. There, as elsewhere, change ruled the roost.

We lived at the foot of Stratton Mountain. Like the other Green Mountains of Vermont at that time, Stratton was covered from foot to peak by a forest of evergreen and deciduous trees. Its 24,000 acres were owned by paper interests who had bought up the mountain and were holding it as a forest reserve. The village of Stratton, a one-time prosperous farming community, had been reduced to a few scattered families. Stratton roads had been neglected or abandoned. Once-productive fields had been taken over by a thriving forest with half a century of untended growth. Neighbors in our valley called the place "the wilderness." It provided them with a source of a few random Christmas trees, balsam fir for wreaths and successive crops of ferns which were picked and sold to florists in Boston and more distant cities.

Across the town road that separated our farm from our neighbors to the south, the land sloped upward toward Stratton Mountain, cutting off the sunlight on short winter days by half-past three in the afternoon. North and east of us, lesser mountains, mostly forested but denuded of big timber, stretched for miles along the banks of the West River.

After we had lived in Vermont for about fifteen years, the paper interests started to cut the forest in preparation for developers who planned a new life for Stratton. Clearings were to be made on the north, east and west slopes. Roads were to be constructed and paved. Almost overnight the Stratton wilderness disappeared. Its place was to be taken by the Stratton ski slopes—one of the largest and best-advertised ski developments on the East Coast.

Our Forest Farm nest became less and less desirable to us as these plans of the ski promoters took shape. The Pikes Falls of the early 1930's was to be swallowed up by the forces centering around a ski-town type of life.

The whole tempo of the valley was changing. Our once-isolated farm had become too available to outsiders from the New York City area and Boston. Countless visitors dropped by, and others moved in. These people were less austere and hard-working than those we had lived with. They were more interested in having a "good time" than they were in hard work. They were vacationers at heart, not workers. The outside world became too much with us. Our seclusion was gone.

Other local circumstances also led us to consider a move from the valley where we had thought we were settled for life. When we came in 1932, we settled down quietly among people who had farmed there for generations. There was little community life, though there was occasional help among neighbors. There were certainly local feuds. By and large it was a stable grouping of families who worked for a living on their own land.

The war changed things. Some of the older folk went to work in war plants; some of the younger were conscripted. Quite a number of conscientious objectors refused to become killers and moved up to our valley. They talked about a good life, about a living from the land, about arts and crafts and even about hard work. We saw the possible dawning of an economic basis for cooperative living. A wholesome, rewarding community life might be built up, we hoped. We endeavored to do our part, on our own place and in the neighborhood.

After war's end, in 1946, there came a change locally. People let down, lowered their ideals, got back to normal. They took things easier, gossiped, feuded again. The talk of community died down. There were cliques and little cooperation. People wanted freedom at all levels—freedom from work, from discipline, from community responsibility. They were out for a good time, for fun, for plenty of leisure to gad about.

Community enterprise, such as it was, in the valley became largely centered around dances and beer parties. There was no community economic enterprise and little social enterprise. People wanted to escape thoughts of wars past and impending. We were told in so many words: "We don't want to hear anything about politics. We want you just as neighbors. Come to the dances." Or, "Come to see us. Pay us a social visit. Have some small talk." And, "Beer promotes good fellowship."

In almost every newcomer's house in the valley, dancing and liquored parties were the social enjoyment of the young people. To what purpose? We felt that life was earnest, that it was an opportunity to learn, to serve, to build truth, beauty, justice into the world. If this were not so, dances, gossip-bees and beer parties might be in order, because then life would be futile and meaningless and any form of escape would be preferable to boredom.

To us, life was real, vital, urgent, important. There were many things that needed doing. We preferred to spend our time and energy in places and among people who were similarly concerned and who were prepared to discipline themselves and organize community affairs in such fashion that the issues of the day were met and the difficulties surmounted. In any case we were not happy in surroundings that were becoming a center for trivial activities and purposeless living.

The straw that really broke the camel's back and made us decide to leave Vermont and seek more isolation elsewhere was an episode that brought a group of seventeen uninvited, unknown visitors. They turned up one working day and trooped into our kitchen on a morning when we were boiling maple syrup into sugar on the stove for a rush order. The people were inoffensive enough—sightseers enjoying an off day in the backwoods observing the natives. For them it was an escape from the treadmill of urban existence. Time hung heavy on their hands, and they were killing time.

"What a lovely house! And you did it all yourselves!" they exclaimed. "And is that how you make maple sugar?" "We want to see the sugar bush." "Do tell us something of your lives." "How did you ever find this place?"

We had chores to do and deadlines to meet. We shooed them

upstairs to look at the view from the balcony. After they had gone, Scott burst out to Helen: "This is becoming impossible! How can we get our work done if we are interrupted every twenty minutes by a new bevy of sightseers? We'll have to move!"

Up to this point Helen had borne the chief burden of dealing with such casual drop-ins. She had greeted them and engaged in the necessary conversation. Scott had stayed in his rock-walled study on the big boulder behind the house. Then she began sharing the burdens of tourism, saying to visitors: "Scott is out in his study on the big rock; why not climb the steps and see what he is up to?" Helen's system brought the expected reaction from Scott. We began to think seriously of moving.

When the facts are set down on paper, they seem inoffensive enough. What if a couple of dozen outsiders a week did upset our routine and seventeen or so descended on us one morning? We could put aside our work for the time being and indulge in conversation, sometimes desultory and sometimes very worth while, with a group or two. But the days had passed into weeks and months of it. All summer long, visitors came and went, until both of us agreed that Pikes Falls, Vermont, had become an impossible place for us and our projects and program.

Where to go? When? And how to go about the search for a continuation of our chosen version of the good life? Where should we find another farm as isolated, as productive and as beautiful as our Vermont Forest Farm?

Our requirements were: isolation enough to avoid the hustling and jostling of the city and its suburbs; a minimum of fertile soil on which to grow our food; abundant fresh water; a woodlot to provide our fuel. And this time, instead of in the mountains, we could look for a place on the water. For nineteen years we had homesteaded in the Green Mountains of Vermont. Now, as we pictured our future, why not spend the next twenty years beside the sea? Both of us enjoyed the hills; both of us were equally attracted by large bodies of water.

We had traveled enough to be able to say that we had seen the world. Where should we go? We agreed in avoiding the Arctic and Antarctica. We agreed in avoiding the tropics and semitropics. With the Arctic and the tropics out of bounds for us, where should we turn? To the temperate zone, of course, but on what spot, on which continent? Europe perhaps? Helen loved her mother's land of Holland and we both had enjoyed time spent in Austria's Tirol, but Europe was crowded, and there was the language difficulty. Everything considered, we decided to stay in the United States where we were born. We would try to find another farm in New England, where we had spent the last twenty years under conditions that were satisfactory to both of us.

One important factor in making our decision was that we could afford to live in New England, particularly in its more remote areas, where land was still available at a reasonable price.

New England has another advantage—its climate. We enjoy the change of seasons and relish each one with its infinite variety. Autumn we appreciate the most, with its crisp weather, the leaves' deep coloring and the absence of bugs. The deep freeze of winter months compels a substantial change of occupation, enabling us to follow our professional and avocational interests.

New England also has geological advantages of considerable importance. Between its minor river valleys, it consists of a succession of secondary mountain outcrops that divide the landmass into relatively small areas of well-drained soil and considerably larger areas of rugged slopes that make open cultivation difficult and emphasize the usefulness of hill pastures. Such broken landmasses lend themselves to the homesteading that has played so significant a role in New England's history.

In making our choice of a New England homesite for a second time, we decided to follow the example of a widely publicized Maine citizen, Henry Gross, and use a dowsing rod to find our future farm. At all times in human memory, as set down in folklore and written records, there have been people with the usual five senses plus a sixth or perhaps a seventh or eighth sense which reports the existence of underground water and even indicates its distance from the surface and the volume of its flow. Water diviners sense vibrations more diverse than those reported by the five senses. Helen is one of these people. When we were looking for water on two farms in Vermont, she had been able to detect its presence.

One day we had run across a book, *Henry Gross and His Dowsing Rod*, written by Kenneth Roberts (Garden City: Doubleday, 1951). Henry Gross, besides finding water for his friends and neighbors, had, while still in Maine, taken a map of Bermuda, and, following Kenneth Roberts' directions, secured fresh water in an area where no water had been known to exist. We decided to use the same technique for finding a farm by the waterside.

Taking a detailed map of Maine, Helen went back and forth, fixing her mind on the kind of place that we wanted and asking the pendulum to indicate that place. Consistently the pendulum circled the Penobscot area, at the head of the bay.

We set aside a week, crossed the 350 miles that separated Jamaica, Vermont, from Penobscot Bay, Maine, and began our search by visiting two organic gardeners who lived very close to the point on the map indicated by Helen's pendulum.

Our organic gardener friends received us hospitably but could not think of any parcel of land that met our specifications. We had supper, stayed the night and in the morning continued our search along the tortuous Maine coastline up toward Canada.

We visited real-estate offices, inspected properties. We even took a boat and looked over land on an island.

We spent five days in the search with no tangible results. On the sixth day we headed back toward Vermont. On the way we passed close to the farm operated by our two organic gardener friends. It was getting on toward evening and we were still far from our Vermont Forest Farm. We stopped once more at their David's Folly farm. They welcomed us cordially and asked what luck we'd had in finding the place we wanted. We told them glumly that we had had no luck at all, had given up the search for the moment and were returning empty-handed to Vermont.

"We've been thinking again of the kind of place you want: a fertile farm, isolation and on the water. We think we know just the place for you, but we don't know if it is still on the market." We contacted the owner, Mary Stackhouse, and yes, the place was still for sale. She had found it too lonely for one woman to live way out on the point of Cape Rosier, and on her desk at that moment was an ad she was sending to the *Boston Globe* in an effort to sell the place right away.

We lost no time the next morning in driving down the miles of narrow dirt roads that led to the Cape. The farm in question was overgrown and neglected, with the meadows in poor condition. On one side of the house the soil was predominantly clay, on the other side sandy loam. There was a bubbling spring on the edge of the woods that ran by gravity into the kitchen. That was a great find. Helen went into the house for a few minutes and satisfied herself that it was sunny and roomy.

The farm was certainly isolated, down on a dead-end road with no near neighbors. It was on a lovely little westward-facing cove of its own. The land could be brought back to good tilth. There was good water. In fifteen minutes the decision to buy was made.

These events took place in the fall of 1951. We were already in contact with a hardware salesman from Hartford, Connecticut, who was eager to buy our maple farm and expected to operate the sugar bush as his source of cash income. We arranged to work with him and his wife and show them the ropes of sugaring during the coming 1952 sap season.

During the late winter of 1951 and early spring of 1952, our pickup truck and a friend's ton vehicle made several trips moving our belongings over the long miles that separated Pikes Falls, Vermont, from Harborside, Maine.

In the spring of 1952 we finished out our last sap season. With keen regret we turned our backs on the most enjoyable work we had ever done in Vermont: tapping maple trees, boiling down the sap and making sugar and syrup from the hard maples that studded our Forest Farm. We were also sorry to leave our sturdy, attractive stone buildings that had been such a pleasure to build and to live in. The hills also we had to leave behind, and a few choice friends with whom we had found much in common.

As the snow melted in the late spring, and planting time beckoned to us, we went up to work over the new-plowed land that was to be our Maine garden for the next quarter-century.

"Now, welcome, somer, with thy sonne softe, That hast this wintres weders overshake."

Chaucer, The Parliament of Fowls, *1380*

"What can your eye desire to see, your ears to hear, your mouth to take, or your nose to smell that is not to be had in a garden?"

William Lawson, A New Orchard and Garden, *1618*

"Spring is the most busy and hurrying season of any in the year."

Samuel Deane, in The New England Farmer, *1790*

"In March it is time for winter to depart, but he may be compared to a crocodile, who, having paid you a visit and staid as long as he ought, pretends to go away; but while he puts his head and body out of doors, leaves his huge tail writhing, bending and brandishing behind. Thus, during March, winter's tail is left to annoy us with squalls, gusts, tempests, rain, hail, snow. There often seems to be a strife between the seasons, spring and winter alternately getting the ascendancy. But, after a while, the latter finds his icicles melting away, and to avoid being reduced to a stream of water, he slowly retreats, first to New England, lingering along the Green Mountains, till pursued by the Genius of Flowers, he goes across Hudson's Bay and hides himself behind the hills of Greenland, till he can venture out again with safety."

Peter Parley's Almanac *for 1836*

"Winter having passed away, the time for labor and the singing of birds again returned. . . . It was now the season for me to bustle about, fix up my land, and get in my crops. . . . All my visitors from the city were surprised to see the garden so free from weeds, while they did not fail to notice that most of the vegetables were extremely thrifty. They did not know that in gardens where the weeds thrive undisturbed, the vegetables never do. As to the neighbors, they came in occasionally to see what the women were doing, but shook their heads when they saw they were merely hoeing up weeds. They said that weeds did no harm, and they might as well attempt to kill all the flies. They had been brought up among weeds, knew all about them, and it was no use trying to get rid of them."

Anonymous, Ten Acres Enough, *1864*

"Warm summer sun, shine friendly here;
Warm western wind, blow kindly here."

Richard Richardson, "To Annette," *1890*

Chapter 2

SPRING AND SUMMER GARDENING

POETS think of spring as the season of showers and flowers. "It is not raining rain to me, it's raining violets." For us in Vermont, spring had meant the delightful task of sugaring. Now we were to adapt to a new kind of season.

In Maine, spring begins with the disappearance of ice and snow in the first real thaw, which may not come till April. Up to that time, zero and sub-zero temperatures may hold the earth in their iron grip for weeks or months on end. But the sun is moving northward and with the lengthening days temperatures climb steadily. Spring is in the air. Spring sunshine alternates with April showers. Sunshine warms the air. Showers melt the snow.

Our first Maine spring was unusually warm. Ice and snow disappeared. We had had the newly marked-out garden site plowed and harrowed the previous fall—for the first and last time. (Since then we have done all cultivation with hand tools.) We had covered the upturned soil with a heavy mulch of hay which we took off in the spring.

As soon as the ground dried out, we spread compost (brought from Vermont in old sap buckets) and worked it lightly into the soil. Day after day the dry weather provided conditions seemingly ideal for planting. We began with onion sets, radish seed, mustard, early lettuce, beets and carrots. In preparing our soil for planting, we had added a protein meal of our own contriving: soybean meal, cottonseed meal, sifted wood ashes, screened peat moss, pulverized phosphate rock and granite dust.

Still the good weather held. We put in our early peas. To our surprise, the next day some of the pea seeds were on top of the ground. We poked them back under. Days passed; the soil grew more dry; nothing much came up. One day our onion sets, planted more than an inch underground, began to appear, lying on the surface of the rows in which they had been planted. That told the story. They had been dug up. No animal that we knew ate onion sets. Birds had obviously scratched through the planted land looking for particles of soybean meal, to which they are very partial. The soil, reduced almost to dust by the dry weather, was easily scratched away by even the smaller birds, who tossed peas and onions aside in their search for soybean fragments or for seeds.

By the time the first showers came, wetting the soil and helping the seeds to germinate, our carefully planted rows had disappeared. Germinating seeds were scattered over the entire garden patch, hither and thither. The bird population in its scratching frenzy had reduced our carefully laid-out truck garden to a shambles. We refertilized (without the soybean meal), reworked and replanted the entire area.

Since that enlightening disaster we have been careful about early planting whenever the topsoil begins to dry out. Fortunately, parched earth early in the New England spring is a rarity. It can happen, however, and when it does occur it will bear watching.

Whether or not you are starting from scratch (as we did in more ways than one in this first Maine garden) in your quest for gardening experience, there are several things that you should do with a piece of ground that you plan to cultivate. It must be cleared and drained, the sod broken up, the ground so fertilized that the balance between nitrogen, phosphorus, potash and more than a dozen other soil elements has been established and can be maintained.

It might be well on new land to visit the county agricultural agent, consult his soil maps and have the soil on your proposed garden spot tested. It will cost little or nothing and will give you a good start on your garden. The county agent will make general suggestions which should be followed until your own experience enables you to take over and make your own decisions.

Then you must work the soil. Working the soil in this sense means sifting it, breaking up big lumps, refining it to the point at which each rootlet on the growing plants can readily secure a correctly balanced ration that will enable it to go on growing. If the chemical balance and the texture of the soil are correct; if light, sunshine and moisture are adequate; if temperatures are in the proper range, your seeds should germinate and your plants should grow, flower and bear fruit. The health, vitality and growth of the vegetation in a particular row of plants depends partly upon the variety of the seed or transplant put into the newly prepared topsoil, but mainly on the balance of the fertilizers.

There is a saying that there is nothing sure in life except death and taxes. We would like to add some other certainties. (1) With almost infallible accuracy a beet seed will produce a beet and the onion seed an onion. (2) If you get the right combination of warmth and moisture, seeds will germinate. A botanist can set down—with almost mathematical accuracy in a sequence of stages—the indescribable drive, push and persistence of sprouting seeds. (3) If the soil in which the seeds are lying is properly provided with a balanced ration of nitrogen, phosphorus, potash and the necessary trace elements, the result will be growth, flowering and fruiting.

Our organic garden is fertilized largely by humus from

compost piles of our own making. We have more than a dozen such piles at various stages of maturity, with several tons of compost in all. In Chapter 6 we will describe the process that provides us with the great bulk of the fertilizer we use in growing all our crops.

Significantly enough, the forest does this job for itself. It is a simple process that proceeds uninterruptedly twenty-four hours each day, from season to season, from year to year, from century to century. Leaves, twigs, branches and sometimes entire trees are shaken loose by rain, wind, frost and snow. They fall to the forest floor and lie layer above layer, enriched by animal droppings, sometimes by the bodies of birds and other animals. In reforesting areas of New England, barring fires which reduce the soil cover to ashes, the soil of an undisturbed forest builds up an inch in about three centuries. The process is slow, but as the volume of humus available for its own further growth increases, it pulls itself up by its own bootstraps.

The gardener must do much the same thing for his plants. With planning and direction, making topsoil ceases to be a process and becomes a policy, with limits set by available materials. Each crop, each day, each growing season, the competent gardener deepens and enriches his topsoil. The forest makes its own compost through centuries; we gardeners can make excellent compost in a few weeks.

Long-extended New England springtime, with its misty days and frosty nights, provides a period of ten or a dozen weeks, from late February into early May, with sufficient daylight and sunshine to encourage hardy greens to push their growing centers farther above the earth without danger of frost damage. Our last possible killing frost hit our Vermont gardens in mid-June. It hit our Maine garden only in late May. By June no more serious frosts were to be expected. We could work over our soil, add our compost and other fertilizers, work them thoroughly into the soil and plant the bulk of our summer garden crops, which comprise corn, beans, squashes, cucumbers and tomatoes.

The general layout of the garden necessitates keeping beds for permanent items such as asparagus, rhubarb, berries and perennial herbs in set areas. Otherwise, the entire garden should be systematically rotated so that the same crop does not get back into the same spot without an interval of at least one growing season.

Generally, in laying out a garden, low-growing plants such as spinach, lettuce, carrots and onions are grouped together. Similar groups are made of tall crops such as pole beans, sweet corn, staked tomatoes and brushed peas.

If the rows or beds of the garden run north and south, rather than east-west, a maximum of direct sunlight will get to the roots of the seedbeds or to the roots of plants in the crucial period of their early growth.

The spring-summer garden is not left to chance. We have a loose-leaf notebook in which we keep current and past records of the garden: row by row, bed by bed or section by section for each year. This includes date of plantings, varieties planted and results for the season. The pages for each year are on file so that we know, from year to year, what crops have been planted and harvested in particular parts of the garden.

Each row or bed is numbered as part of the overall garden plan for the current year, with rotation of crops kept strictly in mind. For records of garden developments, garden rows or beds may be marked with numbered stakes. We make our row- and bed-marking stakes of cedar. They are sixteen inches tall, pointed with a hatchet and numbered with a blue or black lumber pencil.

For family-scale gardening, a few tools only are necessary. Any garden of less than a quarter of an acre (100 x 100 feet) can be worked with a shovel or spade, a hoe, rake and garden string, a supply of marking stakes, a hatchet and a watering can. Our favorite garden tool is a Planet Junior hand cultivator. We are not sure this is still on the market, but some form of hand-pushed cultivator is available and very useful.

Success in spring and summer gardening depends on getting seeds into the earth with a moderate soil cover of at least a quarter of an inch, and keeping them in the soil while they germinate; they must then get roots down and their stems or leaves through the soil surface as speedily as possible. Once this growth process is begun, it should go on unchecked, steadily and rapidly, as the plant builds itself into a self-generating part of the life process.

It goes without saying that, other things being equal, the more fertile the soil the more rapid and more extensive the growth of the plants being fertilized. Of course there is an optimum level beyond which increased amounts of fertilizer do not result in increased quantity and quality of produce. One time when we sent in samples of our garden soil to be analyzed by the experts at the University of Maine in Orono, the reply came back: too rich; cut down on your compost spreading.

Many seed houses print on their seed packets a normal life span for each plant. Oak Leaf or Simpson lettuce, for example, should take forty-five days from planting to maturity. Extra fertility may alter this life span or it may not. The gardener who seeks the best and the most in product must experiment in order to determine the possibilities in any particular case. Generally, if the gardener can afford the outlay, extra fertility pays off in both quality and quantity.

Gardens may be planted in rows or beds. They may also be broadcast. Our preference is for rows in general and beds in particular cases. Tall-growing plants such as pole beans are grown on poles or trellises. In these hurry-up days, gardeners tend to plant dwarf or hill crops and avoid trellises, which are more trouble and take time to set up. We may be in a minority, but we are using more trellises and longer poles.

Years ago we decided to bypass bush beans. It is so easy to put in a row of bush beans and so much more effort to get the poles, place them and do the planting around each pole. Nevertheless, we are convinced that, per square yard of soil, the crop yield on poles is much greater than on bushes.

Thirty inches between rows had been most satisfactory for bush beans. With poles, we began with 3 feet between the rows. With 8-foot poles, 4-foot spacing kept the developing vines largely in shadow. We tried 52 inches between the rows of poles. Today we are inclined to make the distance 5 feet, with the poles 5 feet apart in the rows.

Given good growing weather, pole beans reach the top of an 8-foot pole quite early in the season, then continue to put out vines, flowers and additional beans. We asked ourselves what would happen if the poles were 12 feet instead of 8 feet in height. Experience has convinced us that the average pole bean in an average season will cover a 12-foot pole with foliage, producing both flowers and mature beans all the way up. Today we are experimenting with even taller poles, hoping to get more beans per pole, as we have heretofore.

If the beans are to be picked and eaten green, additional pole height is questionable. One would need a stepladder from which to pick. But if the beans are to be matured and allowed to dry on the poles, there is no picking until the first frost. The bean plants can be uprooted when frost threatens. The vines are left on the pole; the bean pods dry out and are easy to shell.

There is a tendency nowadays to grow dwarf peas. We have compared them with high-growing "telephone" varieties. In our experience the tall-growing peas are superior in both flavor and volume of yield to the dwarf version. Certainly the picking season is longer on the tall peas. Until a tall pea vine is blighted or uprooted, and so long as it continues to flower, it will bear edible peas.

Those who grow tall telephone peas tend to put them on galvanized wire fences. As the sun gets higher, the wire gets hot and stays hot so long as the sun shines. The delicate pea vines like cool weather and resent being burned. Under these conditions we have elected to grow 5- to 6-foot telephone-type peas and to grow them on brush from our own woods.

As we go about our forestry and trim or cut trees, we take a look at each trimmed-off branch. If it is flat and fan-shaped, we lay it in a special brush pile. Periodically, we go over this brush, pick out the likeliest limbs, trim them when necessary, point them with a hatchet and store them off the ground on our pea brush rack. Some of our pea brush is 10 feet high, and the pea vines have often climbed up to the very top.

When our young peas are 4 or 6 inches high, we make holes in the soil with a bar, push the brush into the holes and tap them lightly with the bar. When the whole row is brushed, we take a wheelbarrow of rooted sod and pack it around the pea brush. This steadies them against high winds and provides the peas with a mulch.

At the end of the pea season we sort out the pea brush and put the best on a rack, where it stays until the next pea season. The rejected brush is kept until a dry, hot day, sawed with a pruning saw into 16-inch lengths, packed in paper cartons and stored in the woodshed. It makes excellent faggots for kindling fires.

Early in our gardening experience we learned that fast-growing weeds will often overshadow or crowd out the plants chosen by the gardener to occupy a given row or bed. If weeds get a good start, the future of the garden may be at stake. We know that a current theory is to let everything grow and pick out the edible plants when the time comes. We have never followed this practice.

There are three ways to deal with weeds. One preventive, which we have never practiced, is to lay down building paper and cut a hole where a selected plant is wanted. Another is to go over the bed and pick out every perennial weed by hand. What we do is to hand-cultivate lightly after every wetting, from rain to irrigation. If the cultivation is done carefully, the soil surface will be broken except in the planted rows and germinating weed seeds uprooted. This is a form of early birth control. Periodically the gardener can work out each row, thinning where necessary and removing any weed that has found its way into the seed row. Our vegetable garden covers about a quarter of an acre. If this area is gone over periodically, especially after each wetting, the possibility of weed growth is reduced to a minimum.

Incidental to cultivation, but of vital concern to garden growth, is the practice of side-dressing or mulching. A row or bed of lettuce, for instance, is coming close to maturity. It has grown steadily and well through its entire life cycle. Before cultivation or watering, a light dressing of compost spread close to the lettuce plants and worked in by the hand cultivator can provide the stimulation needed to convert excellent lettuce heads into superb specimens. Side-dressing is not a necessity; the crop in question may be doing very well, but an additional bit of attention may cover the slight spread that separates good from better to best.

Mulch is a layer of litter scattered over a piece of land to prevent the growth of unwanted plants, to check evaporation, to keep the soil cool in warm weather and warm in cold weather, to filter sunshine and modify its extremes. Any light loose material may be used as a mulch. We used branches, seaweed, hay, straw, autumn leaves or other litter. In dry weather a dust mulch checks evaporation. In cold weather a dust mulch protects against undue freezing.

Sunshine and fresh, clean air are essentials of most plant growth. A third essential is water. Plants share with animals this common feature: a large part of their bulk consists of water. Young plants and new transplants are peculiarly dependent on an adequate water supply. Gardens that depend on artificial irrigation perish when the water supply is cut off. Plant growth can be retarded by an undersupply or an oversupply of water.

New England gets enough rain and snow to provide an annual precipitation of about 45 inches. Fortunately, this is moderately well distributed through the average year. To be sure, there are dry periods and wet ones, dry years and wet years, but through more than four decades of New England gardening we have never had a general failure of crops due to too much or too little moisture.

photo: Ralph T. Gardner

"Nothing can be more abounding in usefulness or more attractive in appearance than a well-tilled farm."

Cicero, De Senectute, *45* B.C.

"He certainly is worthy great Praise and Honour, who, possessing a large and barren Demesne, constrains it, by his Industry and Labour, to produce extra ordinary Plenty, not only to his own Profit, but that of the Public also."

Sir Richard Weston, in New England Magazine, *no. 3, 1759*

"In our present imperfect condition, a beneficent Providence has not reserved a moderate success in Agriculture exclusively to the exercise of a high degree of intelligence. His laws have been so kindly framed, that the hand even of uninstructed toil may receive some requital in remunerating harvests; while their utmost fulness can be anticipated only where corporeal efforts are directed by the highest intelligence."

R. L. Allen, The American Farm Book, *1849*

"O suns and skies and clouds of June,
And flowers of June together,
Ye cannot rival for one hour
October's bright blue weather."

Helen Hunt Jackson, Verses, *1884*

"Such Gardens are not made by singing "Oh, how beautiful!" and sitting in the shade."

Rudyard Kipling, The Glory of the Garden, *1911*

"A garden is a work of Art using the materials of Nature."

Anonymous

"The leaves fall early this autumn, in wind. The paired butterflies are already yellow with August."

Ezra Pound, A Letter

Chapter 3

THE FALL GARDEN

In our part of New England, the general gardening practice is to start planting on Decoration Day, which is late in May. Hardy things are planted first, followed weeks later by the more perishable crops. This sequence carries the garden to midsummer, when planting usually stops. Gardening is considered ended for the year in August, except for harvesting. When this is over the land is left fallow, or cover crops are put in or weeds accumulate. Major gardening is considered over till the next spring.

Our practice is quite different. It closely approaches the Japanese way of gardening. Their land is so circumscribed that they must economize drastically on space. When they take out a radish, they replant a lettuce or other seed in the vacated spot. When we take out any section of a bed or row, we do almost the same as the Japanese until well into September. We plant in the spring; we plant in the summer; we plant in the fall. As planting space is opened up by harvesting early summer greens and roots, we immediately put in some other crop that can be planted late and will mature before or during light freezing.

Fall days with us are sunny and crisp, closely approximating the days of early spring in temperature. So we plant in the late summer and early fall the same type of vegetable that flourished in the spring and that again will have time to ripen in the fall: radish, lettuce, chard, mustard, spinach, collards, early cabbage for greens. Even carrots when planted late will mature in the fall into little "finger" delicacies. All of the items we have mentioned thus far are frost hardy. Most of them will live and thrive with night temperatures as low as 18 or 20 degrees Fahrenheit. Some of the seeds will lie dormant and fail to germinate; some will break ground and be frozen out. Many will sprout and grow. The results of fall planting have been well worth the effort, time and our small expense for the experimental seeds.

This means that in September and October, when most other gardens are empty or weed-choked, our garden is full of up-and-coming greens. The fall garden can be almost as green as the spring and summer garden. The Decoration Day to Labor Day gardener does not expect this to happen. Visitors to our fall garden often remark on the amount of vegetables still in the ground. Members of a local Garden Club visited our place one day late in September. There was hardly a square foot of garden space empty. They exclaimed: "Your garden is as green as it was in June. It looks like spring, and we are almost in October. How do you do it?" Our answer is simple: continue planting.

Early in the summer, when the first mustard greens, lettuce, spinach and bunch onions are moving from the garden to the kitchen table, we are busy replacing them with root crops such as turnips and beets, which in turn will give way to young greens for fall use. At the same time in the early fall that hardy greens are going into the ground as seeds, it is possible to transplant main-crop lettuce, endive, Chinese cabbage and celery plants from seed flats to the garden beds. Following this system, the fall vegetable garden can supply fresh greens and roots a couple of months after early frosts will have wiped out squash, beans and tomato vines, ending the spring and summer garden.

Foresight and a few seasons of experience will tell the gardener what to expect and when to make the necessary shift of crops coming out and crops going in. Here is an example:

Early smooth peas are harvested and eaten and the vines are ready to be pulled out of the ground early in July. We replace them immediately by the earliest stages of the fall garden. Even though there are still edible green peas and even pea blossoms on the vines, we agree that the occupied space could be more profitably used. We decide on the day that they should come out.

Instead of the ordinary routine of picking individual peas, we pull out the plants, strip them of all pods and divide them into three containers: the over-mature peas to be dried and stored for winter use; the ripe peas that can go into the day's soup; and the few remaining young peas and green pods that can be put raw in the day's salad.

While this operation is in progress, a member of the garden squad, equipped with a light mattock, loosens the soil where the pea vines were and pulls out any chance weeds. Another member of the garden squad follows with a fork, puts weeds and dried vines into a wheelbarrow and moves them to the compost area, returning with enough compost to put an inch on the area from which the peas and weeds have been removed. A hand cultivator works the compost and a sprinkling of nitrogen meal into the former pea row.

The next operation, with a single-pointed hoe attached to the cultivator, opens a trench along the erstwhile pea row. We sow spinach or some other short-term hardy crop into the row that had been occupied by peas a few hours before.

This operation moves smoothly and is soon completed. All members of the team, whether two or ten, understand what is going on; all approve of it in principle though they may differ as to detail. But if the pea season is to be brought to a close and replaced by spinach or some other short crop, the sooner the peas are out the better. The summer garden has left the area and the fall garden has entered it in less than four hours of a single morning.

Similar operations, repeated as each row is removed, mark the end of the summer garden and its replacement by the fall garden, row by row, with not much loss of time and few unproductive motions.

The fall garden, in terms of its preparation, includes part of July and all of August. In terms of harvesting, it begins in the early autumn and extends through two or three months until night frosts are sufficiently severe to check or even prevent effective growth.

Celery and parsley and spinach will survive moderate freezing in the garden. Broccoli, cauliflower, Chinese cabbage and the hardier western cabbages will take hard freezing. Let them thaw out in the ground on their own roots and then harvest them. If cut for the kitchen while frozen, they will be flabby. If the outside leaves of cabbage or Chinese cabbage are frost damaged, remove the frozen outer leaves. Even after a zero night, such plants may be fresh and entirely edible inside *if* they thaw out before being cut. With south winds and sunny days, the plants' condition will be alleviated by fall and winter thaws that sometimes last for days.

A mulch of hay, straw, leaves and/or evergreen branches laid over crops on especially cold autumn nights can provide an effective cover against the cold. Almost anything from a scattering of autumn leaves over a lettuce plantation to a bed blanket will do wonders to offset frost. We keep a pile of birch, beech or other fairly small leaves handy and dry. If they are scattered over a bed of greens early on a frosty night, they may be effective enough to offset even a 10-degree frost.

Sooner or later, as autumn advances toward chill November's wintry blasts, the garden will lose some of its green freshness and begin to look chilled, but the longer this day can be postponed the longer one can get green nourishment from the fall garden. During the latter part of the autumn-winter gardening period, the adroit gardener can snatch a cauliflower or Chinese cabbage and a hardy chard plant here or there. He can also pick fresh spinach and lettuces and radishes, which have not yet given up their efforts to take advantage of the few hours of growing weather between frost-crusted earth in the midmorning and the onset of sunlessness in midafternoon.

Winter gardeners, willing to experiment, will often have the satisfaction of digging out hardy green plants from under a cover of mulch and light snow. Here in Maine we have had entire rows of lettuce, spinach, parsley and broccoli survive the winter and reappear as the snow covering melts. Brussels sprouts, kale, rape, wheat and rye survive New England winters almost as a matter of course, especially if they are helped along with a light covering of hay, straw or autumn leaves.

When our gardening passed from spring through summer into autumn and winter, we faced a problem that most people would consider insoluble. Cold nights certainly discourage plant growth, flowering and fruiting. We knew this when we began our experiments with cold-weather gardens. We take such adversity for granted. Despite setbacks, however, we have made real progress since the days when 32 degrees was accepted as the point at which most gardeners pick up their tools and go home. Experience convinces us that noteworthy successes are possible with certain hardy plants.

Some present-day lettuces will survive sub-zero temperatures and go on growing and heading. Red radishes have at least a fifty-fifty chance to produce edible roots before they are plunged into permanent winter weather by a degree of frost that keeps the earth frozen hard even on a sunny afternoon. Long after these fall-planted seedlings have given up their struggle against the New England winter, parsnips and oyster plant, witloof chicory, celery roots, and Chinese cabbage and brussels sprouts will survive. Once the ground has been deeply frozen, during subsequent thaws, parsnips, oyster plants and witloof chicory can be dug and turned over to the kitchen.

We have continued our outdoor gardening to later and later periods year after year until heavy frost solidified the earth. We intend to continue our experiments with growing hardy plants in Maine's cool falls and into its cold winters, although we can never be absolutely certain what will survive, particularly with variable weather conditions. Only experience will give us reasonable certainty as a result of these operations. But our efforts to date suggest that we can expect at least a modest degree of success. Each time we eat a fresh salad that comes from our fall-into-winter garden, we have one more assurance that we are on the right track.

photo: Richard Garrett

"The Spring *visiteth not these Quarters so timely.* Summer *imparteth a verie temperate heat, recompencing his slow fostering of the fruits with their kindly ripening.* Autumne *bringeth a somewhat late harvest. In* Winter *we cannot say the Frost and Snow come verie seldome and make a speedie departure."*

Richard Carew, The Survey of Cornwall, *1602*

"I do hold it, in the Royal Ordering of Gardens, there ought to be Gardens for all the months in the Year."

Sir Francis Bacon, Sylva Sylvarum, *1605*

"Snow is beneficial to the ground in winter, as it prevents its freezing to so great a depth as it otherwise would. It guards the winter grain and other vegetables in a considerable degree from the violence of sudden frosts, and from piercing and drying winds. The later snow lies on the ground in spring, the more advantage do grasses and other plants receive from it. Where a bank of snow has lain very late, the grass will sprout, and look green earlier, than in parts of the same field which were sooner bare."

Samuel Deane, The New-England Farmer, *1790*

"I cannot conceive the Spring of lands that have no Winter. I take my Winter gladly, to get Spring as a keen and fresh experience."

Odd Farmwife, The Old Farmhouse, *1913*

"It is said land under glass is fifty times more productive of garden crops than open ground. Glass is certainly the solution of the raw winter greens problem. For with no more than the two-sash hotbed in which we start tomato, pepper and other seedlings, we can eke out the fall lettuce supply until after Christmas."

Henry Tetlow, We Farm for a Hobby and Make It Pay, *1938*

"The pleasure of year-round gardening is one of the greatest tranquilizers there is. In a world filled with tension and frustration, you can enter your greenhouse, close the door, and shut yourself away from all the world's problems. There's something soothing about firming seeds in the soil and tending plants under glass while raindrops and snowflakes fall against the panes."

George and Katy Abraham, Organic Gardening Under Glass, 1975

Chapter 4

WINTERTIME GARDENING

Our comments on fall gardening were written from a homestead that can expect not more than 105 frost-free days per year, with 260 days and nights when frost is possible and/or probable. One of the simplest ways of lengthening the frost-free season for plants is to build a cover with glass or some equally effective transparency, which allows sun rays to enter the area but prevents or retards heat from escaping after sunset.

The problem is complicated by the irregularity of sunlight. If each day were sunny and the gardener had to face only the night chill, the answer would be a simple night covering. New England fall and winter weather includes a minimum of clear sunshine and a maximum of clouds, fog, mist, drizzle, rain and snow. There are periods during the spring and early summer when we do not see the sun once in a week. Such periods may be prolonged in late winter and early spring to weeks when sunshine is a rarity, when day temperatures rise to the thirties and low forties, while night temperatures drop regularly, and often reach the upper twenties. In deep winter, of course, temperatures occasionally fall far below zero.

This need to protect plants from cold weather has produced cold frames (wooden frames glazed by glass or some other transparent material at or near ground level). Using spare time and secondhand materials, any homesteader can provide himself with one or another form of cold frame-greenhouse. Sun-heated glass houses are really just enlarged cold frames. If they are attached to the south or west side of an existing building or wall, in shed form, they are open to the sun, and protected against north winds, and they accumulate whatever warmth the sun is sending. During many days in the winter it gets so warm in our greenhouse between noon and 3 PM that we could take sunbaths there.

Let us begin with the construction of our greenhouse. Our fondness for stonework and the existence in most parts of New England of plenty of fieldstone has made it our choice and custom to build our greenhouses of stone. The present sun-heated greenhouse in Maine is built on a concrete and rock foundation which is 16 inches wide and deep enough to avoid frost-heave. The greenhouse faces south. The north and east walls are stone and concrete from foundation to roof. The south and west walls are made of surplus storm sash. The sloping shed roof is glass, resting on 2 x 4 cedar rafters, and is 9 feet wide and 40 feet long.

The roof is made of double-strength window glass 16 x 24 inches. Both corners of each rafter on one 2-inch side have been replaced by a half-inch groove in which the glass rests. Each pane of glass is lapped one inch on the pane below and held in place by sprigs and glazing compound. The rafters are spaced to allow the glass panes to fit comfortably into the grooves.

The simple shed roof of our Maine greenhouse has a 10-degree slope which is sufficient to drain off water and to be fairly independent of snow and ice. On the outside of the north wall, at a convenient height, we have put a catwalk from which we aim to keep the roof free of snow and ice. During a snowstorm we wait until the internal warmth of the greenhouse has made it possible for the snow to slide easily on the glass;

then with a light wooden pusher we break up the snow cover unit by unit and start the clumps of snow moving toward the low side of the roof. The procedure is possible only when the temperature inside the greenhouse is above freezing.

With concrete foundations and stone and concrete walls that protect on the north and east sides, we find that the daytime temperature inside the greenhouse will be about 20 degrees higher than that of the outside air. On bright sunny days the differential between inside and outside will be much greater—as much as 40 degrees, if no strong wind is blowing. The stones in the north and east walls act as radiators. All day they absorb heat. After sundown they give off this heat as temperatures inside and outside tend to be equalized.

Cloches (small glass enclosures for individual plants), cold frames and formal greenhouses have two disadvantages. First, the initial cost is to be considered. Second, there is always a chance of breakage. Builders of glass houses must take these two items into account when they plan such projects. If plastic greenhouse covers would last for a reasonable number of years, greenhouses could be built far more cheaply. Thus far we ourselves have considered plastic not durable or reliable enough to put into a permanent greenhouse. For the time being, greenhouse builders must shoulder the cost of replacing glass, but with good care and good luck, glass breakage can be held to a minimum, which will more than justify the capital and labor expenses that attend greenhouse construction.

A well-constructed and well-ventilated solar greenhouse will add months to the period during which a New England family can supply itself with greens. Food directly from the garden is unlikely from November to March. This is the period during which the New England garden is least able to furnish fresh vegetables for the family larder and when the greenhouse may be counted on to fill the gap.

The traditional greenhouse provides benches on which pots, flats and other containers can be maintained at waist level, making it more convenient for the gardener. When greenhouse plants are chiefly potted or flatted flowers, the practice may be justified, but in a greenhouse such as ours—which aims to produce many rangy vegetables—it seems desirable to keep the greenhouse floor at ground level, utilizing the full height of the building.

There are three possibilities for the greenhouse floor, presupposing that the greenhouse is a part of the garden area. It may be at the same level as the surface of the garden; it may be above that level; or it may be below it. A pit greenhouse is below garden-surface level. A built-up greenhouse floor is above garden-floor level. Our greenhouse floor is at the same general level as the surrounding garden.

Pit greenhouses have enthusiastic backers who argue that they are warmer and more even in temperature. They cool off more slowly after sundown. It is easier to restore the temperature on each succeeding day because the earth inside the pit is at sub-soil warmth rather than air coolness.

However, dampness and even standing water collects on the pit greenhouse floor unless it is well situated and fully drained. At best a pit greenhouse may be damp. We continue in our opinion that, where there is a choice of greenhouse floor levels, it is easier and more logical to keep the greenhouse floor at about the same level as the surrounding garden than it is to establish and maintain a higher or a lower level.

We go a step further and maintain our sun-heated greenhouse without any assigned walking space. In a 9-foot greenhouse, an 18-inch footway occupies one-sixth of the total floor space. A grower of greenhouse tomatoes, for example, could increase his crop by 16 percent if the path were abolished and gardeners stepped between plants. With us, no part of the greenhouse floor is a regularly trodden path. The roots occupy the entire greenhouse area and with each new crop the footpath shifts.

We built our first glass house in Vermont to give seedlings an early start. Our present glass house in Maine takes care of the seedling problem; it provides us with tomatoes, sweet peppers, eggplants in summer. It also gives us greens during the fall, winter and spring.

Sometime in March, depending on the severity or mildness of the weather, we begin planting seeds in short rows in the greenhouse. Among these seeds may be quick-growing mustard, radish and early cabbage. All of these will be used as salad greens, even when they are quite tiny. Onion, leek and parsley, when well started in the seed rows, are transplanted into garden flats, kept there for a few weeks and, when large enough to transplant again, put directly into the garden.

Flats of young lettuce seedlings will play a leading part in supplying greens in the late winter and early spring. If possible, seeds of the hardiest lettuces should be sowed in September or October in seed flats and be moved into the greenhouse when the weather threatens their further survival outside. These plants, around 2-3 inches high and with sturdy roots, can be set 6 or 7 inches apart in greenhouse flats and/or in greenhouse beds. They should get their start when the greenhouse temperatures are high enough for the plants to get going and become sufficiently accustomed to chilly nights and even chilly days so that they can go on growing when the weather gets colder.

The best lettuces for this purpose are leafy varieties (not head lettuce), with thin central stems and sturdy non-heading leaves. Many lettuces have thick, juicy spines which freeze easily, being moist and fat. We find that the drier the leaf web and stem is, the less damaged by the conversion of water into ice. We tested certain varieties of lettuces that can be expected to survive heavy freezing in an unheated greenhouse from December to March and will begin growing as the sun gets higher and warmer in March-April.

Green Boston, Simpson, Oak Leaf and Buttercrunch lettuce and Celtus (a plant midway between celery and lettuce) suffer comparatively little frost damage. By mid-March, most years, we can start picking individual leaves from all five of these varieties and use them to make salads. We can also strip the outer stems and leaves of parsley and kale plants set in the earth on the greenhouse ground in October. During the coldest parts of winter, growth may be minimal, but the plants survive. We sometimes keep extra garden flats of small plants in the greenhouse all winter to be used as replacements for any transplants in the greenhouse rows that do not make it.

The winter greenhouse also accommodates some adult plants which submit to the humiliating process of heeling. Heeling-in is a procedure whereby the plant is dug up and then replanted with undamaged roots. If carefully done, the plant should remain as upstanding and colorful as any normal plant of its type and age left in the original ground, though there is little further growth.

We might illustrate the process with an example of a bed of giant leeks that were planted in the greenhouse last March, taken out to the garden, and brought back in November, and which we are now eating a year later, in the spring of 1978. They spent a summer of growth during 1977 in the garden and weathered the early frosts of September-October, reaching the first week of November in a state of excellent health, when we decided to bring them under cover.

We picked a cloudy, damp day so that the earth around the leeks would be moist and cling to the roots. We dug a trench 8 inches wide and 6 inches deep in the ground of the greenhouse. Then we went to the outside leek bed with a shovel and used it to pick up a leek plant, being careful to damage its roots as little as possible and to keep an earth ball around the roots. We carried the earth ball containing the leek plant on a shovel into the greenhouse and planted it in the newly dug trench, pulling the earth up around the leek and tramping it in with the heel, pressing the earth firmly.

Then back to the leek bed with the shovel, where we took up a second leek, placing it as close as possible to the first heeled-in leek. We repeated the operation until the trench was filled. With room for a second dozen, we put in more in the same fashion. When the move was completed we watered the leeks moderately. On that day and during the following days and weeks, not a single leek plant wilted or looked or acted any worse for the move. Those that survived our picking through the winter were still green and sturdy in the greenhouse in the middle of April 1978.

We have also dug up and heeled in our parsley plants with hundred-percent survival. We do somewhat less well with lettuce transplanted from garden flats into the greenhouse, but at least half of them survive.

We had even less success with celery. They seemingly cannot take it. During the summer of 1975 we raised from seed several rows of fine celery plants. We had about two dozen summer Pascal plants which had surged to a height of nearly three feet and to great bulk. Some of the plants weighed up to five pounds each. We dug these plants and heeled them into the greenhouse in early November. They took the shift rather badly. Despite plentiful watering, they never fully recovered the upstanding vigor of their celery bed outside, but they did stay alive. Their leaves and stalks were edible until February, when they gave up the fight and collapsed under a fierce bout of sub-zero weather. Meanwhile we had been eating excellent celery from the heeled-in plants for three months.

The fall and winter gardener cannot bother with semi-hardy plants that are rubbed out by temperatures between 20 and 30 degrees Fahrenheit. Experience in each location will show which varieties will survive and which will not survive cold weather. We found that mature chard plants froze after about sixty winter days. Chinese cabbage lasted another month. Parsley and kale plants survived cold winters in the greenhouse, resuming their growth as the outside temperature rose in the spring.

As experiments go forward and hardy new varieties of established plants are devéloped and introduced by selection and crossing strains, the plants available to fall and winter greenhouse gardeners will increase in both number and hardiness.

For those who have the money and the facilities, the possibility of gardening in heated greenhouses is always present and can be carried forward as long as the electric power is on. Here we are concerning ourselves with fall and winter gardening in solar-heated greenhouses. We are among those homesteaders and small-scale gardeners who prefer to have sun-heated greenhouses, to collaborate with Mother Nature, rather than having to deal with the electric and fossil fuel companies.

photo: Gilbert E. Friedberg

"What cost to good husband is any of this?
Good household provision onely it is.
Of other the like, I doo leave out a menie,
That costeth the husbandman never a penie."

Thomas Tusser, Five Hundreth Pointes of Good Husbandrie, *1557*

"In a house where there is plenty, supper is soon cooked."

Miguel de Cervantes, Don Quixote, *1605*

"Gather at the full moone for keeping apples. Let them sweat for ten days on straw. Then pack in bran."

Anonymous, West Country Herbal, *1631*

"The infinite conveniences of what a well-stor'd garden and cellar affords... All so near at hand, readily drest, and of so easie digestion, as neither to offend the brain, or dull the senses."

John Evelyn, Acetaria, *1699*

"There can be no doubt whatever that by far the most economical plan of supplying a household with necessaries for consumption is to lay in a stock for the week or month in lieu of purchasing, as but too many do, from hour to hour."

One Who Makes Ends Meet, Economy for the Single and Married, *1845*

Chapter 5

WINTER STORAGE

THE DANGER of frost for food crops in our part of Maine is considerable. For more than half the year commodities subject to frost danger must be stored or otherwise protected. In a very real sense the success of each homesteading year depends upon carrying perishable food from the period of its production to the time of its consumption.

The methods of keeping some vegetables fresh in the winter greenhouse have just been described. An even easier method of food storage is to leave the hardiest of vegetables in the ground. Parsnips and Jerusalem artichokes, for example, winter in that way better than any other we know of. They may be dug and eaten during any winter thaw sufficiently deep to allow the roots to be lifted from the ground. They may also be dug before freeze-up in the early winter; then they are not so sweet. Unfrozen, the parsnip is as bland as the average turnip. Freezing converts their starch into sugar, making the root more tasty. This is true of both Jerusalem artichokes and the parsnip.

Parsnip and artichoke devotees have a period of perhaps a month, beginning with the first spring thaw, during which the main root may be dug and eaten raw or prepared and cooked to taste. As spring advances, the plants begin to send out small feeders from the main root, which gets more fibrous and less sweet. Under ordinary Maine winter conditions, it is best to dig parsnips and artichokes to be eaten promptly if the upper layer of topsoil is frost-free. During winter thaws they may be chewed by rodents or other animals; in very severe winters they may even be frozen out.

Salsify may be left in the ground over winter. Like the parsnip, the salsify root is most edible immediately after winter break-up and before the new crop of feeding roots has begun to put in an appearance.

Almost all of the roots grown in the ordinary home garden, except potatoes, will survive moderate frosts. Some are improved by moderate freezing. Most of them will be seriously damaged by a series of severe frosts unless well mulched.

It behooves the Maine homesteader to get his or her food directly from the garden as late as possible in the autumn and as early as possible in the spring. Under the best of conditions, winter's deep-freeze hiatus of garden-fresh vegetables will last for perhaps three months. During this frozen-up period in Maine, storage of some kind is essential if roots are to be kept in prime condition.

Most family homes in New England are equipped with some sort of cellar. If the cellar contains a furnace or other heat source, the entire area will be too warm for the storage of perishables. Even if the heating unit is separated from the remainder of the cellar by a concrete wall, the cellar tends to be out of bounds for storage purposes.

Professional storage facilities are equipped with air conditioners that maintain an even level of temperature, winter and summer. Like all mechanical sources of assistance, air conditioners cost care and attention to maintain. In the long run they wear out, deteriorate, require professional attention. Eventually they must be repaired or replaced. Short of a professionally built and equipped storage unit, the problem of keeping fruit and vegetables in winter, like so many other problems, looms on the horizon for the homesteader with the oncome of winter.

The house we bought in Harborside had no central heating, but it did have a small cellar, with sturdy stone foundation walls and a cement floor. In this cellar the temperature seldom fell below freezing and seldom went higher than 45 degrees Fahrenheit. In this cellar we have kept vegetables and fruit from the harvest time of October-November until the following July. Only the best specimens could maintain themselves for months after being separated from Mother Earth. On rare occasions, as with the summer crop of 1975, some of our stored potatoes, onions, turnips, beets, celery roots, carrots, apples and rutabagas were still edible when the new crop was harvested in the late summer of 1976.

It goes without saying that freshly harvested roots are more juicy and tasty than those that have lived in storage during an entire winter season. Stored roots are alive and subject to aging and spoilage. Even under professional storage conditions which hold temperatures at exact levels month after month, losses due to rotting and wilting are expected. We go over the boxes of vegetables and fruit stored in the vegetable cellar once every month or six weeks, pick out those that have turned bad and get rid of them, and use up at once those that show signs of incipient decay.

One of our most valuable stored items in Maine, as in Vermont, is apples. At harvest time we sort our apples carefully into two lots: those that are perfect and will keep; and the non-keepers, which have blemishes or bruises. The bulk of the keepers we put away in autumn leaves, while some we eat or give away. The non-keepers go into apple juice or applesauce.

Our simple recipe for applesauce is to wash the apples, cut the smaller ones into quarters, the larger into eighths (taking out the cores and any blemish or rot), put them in kettles with a minimum of water and cook until they soften up. We then pack the fruit in sterilized jars without processing them further. We seal the filled bottles and store them in the cellar. No sugar or other preservative has been added. When the applesauce is eaten, if desired it may be sweetened with a little honey or maple syrup.

This is an "open kettle" method—the simplest of the simple. Only occasionally one jar in perhaps fifty thus treated will ferment or mould and go to the compost pile. If this happens, it is usually due to a cracked top of faulty rubber. At least 95 percent of the canned material thus stored in our cold cellar is perfectly good when used even after two years.

Our canning of other produce is as easy and casual. When tomatoes are plentiful, we pick them by the bushel and make a soup stock, with celery, parsley and onions chopped up and boiled together; no water is added. These are all soft crops that would not last in the garden through the first heavy freeze, or, as to the onions, they are blemished or too small to keep. The four vegetables are boiled together by the same open kettle method as the applesauce. When the celery (the toughest of the vegetables) is forkably tender, it is all put through a sieve, with the resulting juice being brought back to a boil and then canned. The thick residue that does not go through the sieve is put back on the stove with a minimum of hot water, brought back to a rolling boil and canned with a bit of sea salt in quart jars, to be used later as soup stock.

This is about the extent of our canning, except that we put up a few dozen bottles each of blueberries, raspberries and rose hips, in the following manner. We sterilize the quart bottles, pour in a cup or two of boiling water, add a large spoonful of honey, stir to dissolve, drop in a cup and a half of one or another fruit, add boiling water to the top of the jar, seal and put in the cellar.

We were given a freezer and find it very handy for storing overstocks or surpluses from meals or from the garden. We dispose of masses of raw blueberries in cellophane bags. We blanch extra asparagus and eat the icy stalks in the wintertime without further cooking them (they become flabby and less tasty when cooked). We could live happily without electricity for lighting, but we find the freezer of considerable assistance in keeping certain few foods.

In the absence of a reasonably cool and dry house cellar or a freezer, it may be economical to build a root cellar for storage purposes. It must be set high enough and dry enough and cool enough. Like any other storehouse, it must be proof against invading insects such as ants and against rodents such as rats, mice, squirrels, raccoons and skunks.

One of the early means of storage was a pit, lined and covered with hay or straw, and held in place by boards and loose earth and properly ventilated. Located on a knoll to obviate the accumulation of water, such storage pits proved to be reasonably frost-free. Periodically, on warm days, pits could be opened up, a portion of the contents removed for immediate use and the pit reclosed.

An alternative was to pick a side hill location close to the farm buildings, dig a hole into the hill, build a front with a door and ventilator. The side hill location, well chosen, took care of water accumulation. The part of the storage cellar protruding from the hillside should be double boarded or in some way protected against summer, spring and autumn warmth. The chief obstacle to any of these storage structures is the difficulty of making them vermin-proof.

We will never forget one of our winters in Harborside when we had stored twelve lugs of apples in the cellar supposed to be rat- and mouse-proof. The apples were layered with abundant semi-dry autumn leaves below and above the fruit and a good covering of leaves. We closed the door of the cellar and went off on a transcontinental lecture tour. On our return, four months later, we went to the cellar to look over our dozen boxes of carefully stored apples.

The cellar floor was covered with debris, inches deep. We felt in the boxes. Of the hundreds of stored apples, only one apple remained—a fine specimen of Northern Spy. We never saw the marauder (probably a rat or a squirrel), nor could we find out where it got in. It had gone from box to box, chewing up the apples and leaving plenty of the chewed pulp as a token for us latecomers. What a winter it must have had!

Our neighbor, Eliot Coleman, who maintained an organic market garden next door to us for several years, built a carefully planned root cellar as an independent unit. The cellar consisted of two rooms drained and insulated. It was equipped with double doors and was designed to be vermin-proof. The elaborate structure was of double wood with concrete floor and retaining walls, and, so far as we know, functioned well.

Thus far we have discussed winter storage of living objects like carrots, potatoes or apples. Another method of storage is to dry garden produce sufficiently to prevent its decay. In earlier days much fruit—apples, pears, plums, cherries, and various berries—was dried and stored for winter use. We have not done much of this, preferring to eat the fruits as they come along in season, or canning those that are going by.

Many herbs are dried and stored for flavoring, seasoning, preserving, making herb teas or for medicinal purposes. We have grown various species in our herb garden: mint, thyme, chives, lemon balm, lovage, tarragon, summer and winter savory, dill, marjoram, camomile, coriander. These and others are picked by the branch just before ripening, hung on the kitchen rafters to dry, and used for morning teas or put in soups or salads.

It goes without saying that if possible we prefer to grow all our own food and have it garden-fresh, but in our climate this is not possible. We therefore have worked out these various ways to store, dry, freeze, can and otherwise preserve what foods are not eaten straight from the garden.

photo: Alex Crosby

"Lay durt upon heapes, some profit it reapes."

Thomas Tusser, Five Hundreth Pointes of Good Husbandrie, *1557*

"It were good to trie whether Leaves of Trees swept together, with some Chalke and Dung mixed, to give them more Heart, would not make a good Compost; for there is nothing lost so much as Leaves of Trees."

Francis Bacon, Sylva Sylvarum, *1605*

"Lay your material in a large heap, in some convenient place: A layer of fresh and natural Earth, taken from the Surface, and another of dung, a pretty deal thicker; then a layer of Earth again, and so successively, mingling a load of lime to every ten loads of dung, will make an admirable Compost, somewhat shaded, so as neither the Sun too much draw from it, nor the violent rains too much dilute it."

John Rose, The English Vineyard Vindicated, *1675*

"Have allways ready prepar'd several Composts, mixed with natural pasture earth, a little loamy: skreene the mould, and mingle it discreetely with rotten Cow-dung; not suffering it to abide in heapes too long, but be frequently turning and stirring it, nor let weedes grow on it; and that it may be moist and sweete, and not wash away the salts, it were best kept and prepared in some large pit, or hollow place which has a hard bottom and in the shade."

John Evelyn, Directions for the Gardiner, *1687*

"There being nothing so proper for Sallet Herbs and other Edule Plants, as the Genial and Natural Mould, impregnate, and enrich'd with well-digested Compost."

John Evelyn, Acetaria, *1699*

"The leaf-harvest is one of importance to the farmer if he will but avail himself of it. A calm day or two spent in this business will enable him to put together a large pile of these fallen leaves. . . . Gardeners prize highly a compost made in part of decomposed leaves. The leaf-harvest is the last harvest of the year, and should be thoroughly attended to at the proper time."

Anonymous, Ten Acres Enough, *1864*

Chapter 6

BUILDING THE SOIL WITH COMPOST

Soil is that portion of the earth's crust which supports the growth of vegetation. Soil—plus water, plus air, plus sunshine, plus the magnetic field—provides the medium in which the life forces functioning through plants, men and other self-motivating creatures thrive and multiply.

Plants live on the soil. Animals, including man, live directly or indirectly on plants. Directly or indirectly, the soil (whether topsoil or the lower layers of subsoil) provides the mineral elements of which plant bodies and animal bodies are composed. Sand, gravel and rock particles plus organic matter make up both topsoil and subsoil.

Plant nutrients exist mainly in the topsoil. Through its feeding roots, the plant secures the water and building material which constitute the plant body. Anchor roots penetrate the subsoil sufficiently to provide the anchorage without which the tiny feeding roots would be unable to keep the contacts through which they receive moisture and nutriment.

Left to her own devices, Mother Nature builds topsoil, enriches it with organic material, adds the products of water and wind erosion and thus deepens the layer of topsoil and lays the foundation for grass and later for forest cover. Growing vegetation adds gradually to the means of its own perpetuation and enrichment.

We know that Maine's combination of soil fertility and climate will produce magnificent tree crops. Fine forests were found here by the Europeans when they began prospecting and colonizing the New England area. But colonists who moved from Old England to New England were not concerned with conservation. On the contrary, they felt they had to hustle in order to prevent the forces of nature which dominated life in their newly acquired land from pushing in and obliterating the relatively feeble impact of human handiwork. There was small chance then that man would obliterate nature. It seemed probable that the forces of nature would push man aside and continue with their natural processes.

Maine farmers occupy a segment of the New England countryside that until recent years was practically covered by a mixed stand of softwoods and hardwoods which in its long occupancy of the land had built up a topsoil and subsoil well adapted to provide a livelihood for homesteaders and other would-be occupiers.

In the initial rush to use up and profit by the treasure that nature had wrapped up in Maine's topsoil, the original settlers generally interfered little with basic natural processes. Then, as the population increased and entrepreneurs were looking for a quick profit, they began to slash down the forests wholesale, broke up the grassy covers and opened the way for the grazing, overcropping, erosion and eventual fertility exhaustion that now is so widespread.

With the advances in growing awareness (local, state and national) of the desirability of maintaining present soils and deepening them in the present and future, an entire generation of conservation-conscious Maine youngsters could dedicate itself to building up the Maine topsoil. Every landowner in Maine who has the economic and ecological welfare of the state at heart could follow a policy of putting more humus into the soil each year than is being subtracted by cropping or lost by erosion. Granted, such processes are so slow that a single generation may be unable to observe the losses and measure them against the gains.

Even though quick profits have been made by lumbering, and incidental advantage has been taken of Maine's vacationing possibility, close attention could be devoted to harvesting successive forest crops, and the bulk of Maine's land could be turned to good advantage as arable land.

Certainly Maine homesteaders have little or no hope to "strike it rich" as prospectors did in the early days in California and more recently in Alaska. Instead, they can work out a program that may give them a simple livelihood in exchange for a minimum of consistent and persistent labor. The surest path to success with their land is to increase its fertility.

Homesteaders are often inexperienced and impatient. They are eager for quick returns and will not wait for years to get results. Properly equipped and with a minimum of courage and good health, they can win sure and decisive successes. We know this from our own experience, and some of our neighbor homesteaders know it from experiment and experience.

In Maine an average land-based family of four or five adults and children requires at least a quarter-acre (100 x 100 feet) of moderately fertile garden soil to produce a minimum amount of vegetables, small fruit and greens. The fertility of this garden land can be increased from year to year by deepening the topsoil and adding decaying organic matter (mature compost) and necessary quantities of missing trace fertilizing elements. This land could be used actively for at least eight months out of the year in growing garden edibles. If some glass or other cover was used, a part of the garden space could be used for a full twelve months.

Each year, with proper attention, the topsoil should increase in depth as a result of the application of compost and cultivation. Each year an alkaline balance should be maintained in the soil. Each year any underlying clay bank should be broken up by the addition of compost, sand, sawdust, autumn leaves, wood dirt from the forest floor and peat moss. Each year more nitrogen, phosphorus, potash and trace minerals must be added to the soil than are removed by cropping and erosion, thus adding fertility.

On our Maine farm we are adding our mite to the growth process by increasing the productive capacity of our particular cranny of the earth's surface. Survival and growth depend upon the capacity of the farmer to do a bit more each day and each season. This means a more effective use of the land strip that he has undertaken to cultivate and improve. To do this, the soil must become more productive. As each crop is harvested in our garden, before we replant the area we spread an inch or two of compost and work it into the soil surface; we then plant the next crop. As we are building up our soil, we try to add a bit more than we take out in crops, plus some allowance for erosion.

Compost is decaying organic matter that has reached a stage in its disintegration at which it is ideal food to be built into roots, stems, branches, leaves, flowers and fruits, which make up the physical structure of plants. A good general rule for composting is: utilize any and all available organic materials. Utilize everything that has grown, is growing or will grow. Destroy little or nothing by burning or throwing in the dump.

We visited an organic garden in northern Holland. The husband managed a business; the wife made herself responsible for the garden and did much of the work in it. When we visited their place, one of our first questions was: Where is your compost pile? Our hostess showed us the pile. It was in moderately good order and was made up of kitchen wastes and an abundance of soft annual weeds, with masses of chickweed and purslane.

We asked whether all of the garden wastes went into this pile, as it was rather small for the size of her garden. Our hostess answered: "No indeed! We divide our garden weeds and wastes into two categories: the soft ones that will make compost and the hard ones that will not make compost."

We asked what had happened to the uncompostable materials. "Oh, those," she said, "are down in a heap in the lower end of the garden." In a low-lying corner a place had been used for many years under the direction of this gardener, who year by year had dumped the hard or "uncompostable" weeds in a ridge at least twenty feet long and several feet wide and high.

The ridge was overrun by a jungle of brambles and vines, among which were some nettle plants that were five to six feet high, still at the plant stage but reaching out like embryo trees. We pulled out some of the coarser weeds and dug into the ridge with a shovel. The soil was as black as the proverbial hat. The ten-year accumulation of weeds "too tough to compost" had given our hostess an estimated forty tons of the finest compost that you could hope to see.

Another instance of a real find in organic material happened in Vermont. Our sugar grove had included many weed trees, especially evergreens: spruce, fir and hemlock. Each year in the off-season we cut out these weed trees, limbed them and piled the brush in low spots. As we went back over our clearings, we noticed by the third year after such brush piles had been heaped up and covered by successive layers of autumn leaves, if we turned over the piles of brush and leaves we found feeding roots from the surrounding maple trees working their way up into the rotting brush piles. The same process can be observed in any forest where brush is gathered up and tramped into compact piles. Nature was making compost for the maple trees.

Our chief sources of compostable material are grass clippings and hay cut once a year on our mowings; garden wastes, including weeds, leaves, stems and roots and thinnings; seaweed; grass from salt meadows; sawdust (in small quantities, 5 percent or less of the total pile); and kitchen garbage. We aim to keep reserves of any or all of these materials (except the kitchen wastes) in special heaps or bins near the composting area so that they will be handy for compost making.

Each compost pile should be located so that it can be turned at the proper time with minimum effort. It should be set high enough to avoid standing in water. The compost area should be out of direct sunlight if possible. If the weather is hot and dry, the piles should be watered and covered.

Currently we maintain about a dozen compost piles, 6 feet square, standing in a row and spaced 18 inches apart. Our piles are edged with poles about 2 or 3 inches in diameter and exactly 6 feet long. When we are cutting or thinning in our woodlot and run across a piece that is straight and at least 72 inches long, we say, "That is for a compost pole." We measure it accurately and limb it carefully. Then we put the pole in our reserve pile for later use.

When we get an accumulation of compostable materials and an open space in our compost yard, we take a bar and an axe and start a new pile 18 inches from its neighbors. If there is a grass sod, the area should be desodded so that earthworms can easily enter the pile from below.

We select four good-size, fairly straight poles and lay them in an open square. Across this square at the center we lay four or five slim poles an inch in diameter. They may be left loose or tied in a flexible bundle. This is our horizontal air drain. At the center of the open square we set up four or five vertical poles of the same size, pointed and driven lightly into the ground and tied together with wire or string. This provides for the vertical air drain.

We are now ready for four or five inches of coarse material: weeds, grass, hay, straw, cornstalks—the coarser the material the better, because it allows air and moisture drainage at the bottom of the pile. We spread over it a one-inch layer of good topsoil or compost. This material will contain the insect and bacterial life that plays so important a role in breaking down the cellulose in the pile. There is difference of opinion as to how thick this soil layer should be. We aim to make it one inch.

At this point we cover the earth layer with a thin sprinkling of pulverized limestone or sifted wood ashes if we are making an alkaline pile; for an acid pile we sprinkle ground phosphate rock.

The first layer of the pile is now completed. We next spread three or four inches of seaweed, which we cover with an inch of topsoil, followed by a sprinkling of lime or phosphate rock and then by an inch of sawdust, completing the second layer of the pile.

Each layer of material is held in place by another four compost poles, laid on top of the earlier ones. The poles need not be notched. Early in our composting experience we notched the poles so they would not roll out of place. That made four notches on the end of each pole. In a pile of 40 poles, that was 160 notches. "Why notch?" someone asked. "The poles won't move if other poles are on top of them and if there is composting material to hold the poles apart." We tried the experiment and the poles did not move unless they were pushed by man or animal. From that day on we have done no more notching.

We did encounter another difficulty. If the poles were put in without notches and if the composting materials were spread across the pile and trampled hard, enough expansion was generated by the hard tramping so that, come freezing weather, the composted material expanded to force the poles out of the sides of the compost piles. We stopped tramping and had no further difficulties with this type of expansion.

If garden or kitchen wastes are available, they make the bulk of the third layer. So the pile grows larger, layer by layer, like a layer cake, with an inch of topsoil between each two layers. We go on building as long as it is convenient to reach up and across with hand tools.

If the pile reaches a height of 5½ feet, within three weeks in warm or hot weather fermentation has developed an internal temperature of about 150 degrees Fahrenheit. This heat speeds the breakdown of the pile. At the end of three weeks we cover the pile with four or five inches of hay and let it "cook" for approximately a month.

At the end of the month we remove the hay cover and, with a fork and/or shovel, turn the pile into an adjoining open space, taking off the compost poles as we work down the pile and using them for building the new pile. We turn the corners and edges of the old pile into the center of the new pile and spread the material evenly as it goes in layer by layer. The result will be a heap of semi-decayed organic matter.

We cover the turned pile with hay or straw and let it continue to ripen until the materials can be easily broken up with a shovel. The compost is then ready for use as a fertilizer. Through the years our compost has been the only bulk fertilizer used. We do not touch the bagged commercial fertilizer commonly sold to farmers. Incidently, during this period we have used no animal manures or animal residues such as bone meal or blood meal. As vegetarians, we do not want any part in the raising, exploitation or slaughtering of animals for food.

Given ordinary summer weather, including fogs and rain, the pile should be ready for use in the garden in about two months from the time it was built. On this basis, a six-foot-square pile in the compost yard could produce at least two and often three tons of ripened compost each season.

Our method of composting is a modification of one commonly used by organic gardeners in the United States and Europe. We have been using this method of composting in our present garden for twenty-five years. What began as a tough yellow clay soil that hardened to brick-like consistency if it was worked while the ground was still wet could now pass for a high-level sandy loam soil. Now immediately after a shower we can work the land without having it stick to the tools.

When we started with our garden plot in 1951 it had been a gone-to-seed mowing for years. The chap who gave it the initial and only plowing, our neighbor Russell Redman, commented as he labored over it, "You'll never get a garden on this poor land." He now comes nearly every summer and looks over the garden wall with pride, admiring our produce.

This transformation has been achieved by generous additions of compost, sand, sawdust, and rock powder, plus much tillage and deep spading. Our garden is a living demonstration of the changes that may be brought by using organic materials on a piece of unpromising soil. The time element—twenty-five years of patient work—is also a factor.

"Whosoever will build a mansion place or house, he must situate and set it there where he must be sure to have both water and wood. For profit and health of his body, he must dwell at elbow room, having water and wood annexed to his place or house; for if he be destitute of any of the principal, that is to say, first, of water for to wash and wring, wood to cooke and brew, it were a great discommodious thing. And better it were to lack wood than water, although that wood is a necessary thing, not only for fuel, but also for other urgent causes, especially concerning building and repairs."

Andrew Boorde, A Dyetary of Helth, *1542*

"They are happy, who have a piece of standing water in their garden, or a rivulet near at hand, from whence the garden may be watered without much labour."

Samuel Deane, in The New England Farmer, *1790*

"He who digs a well, constructs a stone fountain, plants a grove of trees by the roadside, plants an orchard, builds a durable house, reclaims a swamp, or so much as puts a stone seat by the wayside, makes the land so far lovely and desirable, makes a fortune which he cannot carry away with him, but which is useful to his country long afterwards."

Ralph Waldo Emerson, Farming, *1858*

CHAPTER 7

WATER FOR HOUSE, GARDEN, AND A POND

PEOPLE who say, "I would like to buy a farm," usually have in mind land rather than water. Yet land without water is all but useless. Whether they are thinking of themselves and their family, their farm livestock, their growing crops or their own hour-to-hour and day-to-day needs, they must include water among their basic necessaries. In homesteading the two prime requisites are enough land and an abundance of unpolluted water.

On the first visit to our Maine farm we asked about water on the place. Mary Stackhouse, the owner, triumphantly took us up to a bubbling spring in the woods about seven feet higher than the kitchen floor, which meant that spring water ran by gravity from the spring into the house. Flowing water, clear, cold and constant, with gravity feed! What more could we ask.

Still, it would be nice to have a pond, for swimming in summer, for ice skating in winter and as a source of irrigation for the garden. Was there any such possibility on the place? We learned that the westward-sloping hillside on which the farm was located had a number of places in which spring grass or water grass grew, indicating the presence of water near the soil surface. The slope also included a swamp area—a place of low-lying land on which water accumulated and stood for much of the year.

How does one go about converting an acre or so of swampland into an acre of pond? The typical American way is to call up the nearest contractor who has a bulldozer and get him to take on the job, finishing it in a few days (and leaving a mess of unsightly piles of topsoil, subsoil, rocks and roots all mixed up together). We had an alternative method. We proposed to do the job with hand tools.

The pond area had not been mowed or otherwise farmed for a generation at least. Filled at the outset with black alder, pin cherry, poplar and willow, the land was rapidly developing a stand of white birch, balsam fir, some spruce and swamp growth. Our first job was to get rid of this adolescent jungle, beginning with the trees. A bulldozer would have taken care of the young trees in short order. We had another way.

The subsoil of our pond area was yellow clay—in places, several feet of it. Such a subsoil repels the anchor roots of trees, so instead of growing down into the clay and hardpan, the anchor roots radiate horizontally on top of the clay bed. We therefore began operations by stripping the sod around a sizable tree and with a grubbing axe chopped through all the side roots. Small trees could then be pushed over by hand, the root cut off and the trunk converted into small wood for the kitchen stove.

Larger trees were treated differently. With a mattock and a grubbing axe we dug a small circular ditch around each tree, cutting off the horizontal roots a foot or so from the base. We then cut the tree about two feet above ground level, pushed a crowbar under the stump at the clay level and lifted the bar vigorously. Three times out of four the tree stump was pulled loose from its clay subsoil and could be tossed into our brush pile.

In an operation of this kind, instead of burning brush and stumps, we moved them into a low spot and filled it well above the level of the surrounding land. In the course of a few years, rot, gravitation and the pressure of winter snow and ice pushed the forest refuse down to the surrounding ground level, and a few wheelbarrow loads of clay or other subsoil held them in place. Thus a single operation disposed of the forest wastes without burning them and raised low-lying land areas above swamp level.

In the deeper parts of the pond we were removing at least two or three feet of soil. When this job was completed the trees or tree stumps were standing high and dry on pedestals that could be removed by mattocks and shovels without serious trouble.

We excavated the pond bed with contractors' wheelbarrows with 16-inch wheels and pneumatic rubber tires. When the area was wet we used loose, movable planks for wheelways. When it was dry we could wheel on the clay.

Our first cargo from the pond area would be sods—not polite, neat lawn sods but masses of vegetation—roots, leaves and branches. Coarse vegetation, root and leafage, went into sod piles built, like our compost piles, log-cabin fashion, with poles. Topsoil went to the gardens to be used for compost making or to level up garden surfaces.

From the outset this dam had a plan, and a purpose. Each of the 16,000 wheelbarrow loads taken out of the pond during the years was aimed in a certain generally recognized direction: a shallow pond with a spillway and an emergency overflow, a body of water that could be siphoned into the garden and that reduces fire hazard in the dry season and provides excellent skating in winter.

At odd times on this dam-building project as many as a dozen people were working at the same time. They had never all been together before and probably never would be again. They were of different ages, sexes, races, political and religious beliefs. They had all come because they wanted to and stayed for the same reason. With rare exceptions, they were not being paid for their work. If they stayed until noon they would share in the same simple, abundant vegetarian lunch.

Relations with the pond were not always successful. One negative item was leaks. At various times serious openings appeared under the 40-foot-long concrete core we built in the 20-foot-wide earth dam, and around the 8 x 10-foot concrete block that provides our main spillway and through which runs the bottom-level tile which drains the pond.

Why should such leaks occur after we had taken the greatest possible care to make it all water-tight? Possible explanations are legion: a mouse run, a muskrat burrow, a frost crack in the concrete, the various rates of expansion and contraction of the concrete and the earth surrounding it. Perhaps even so small a break as an earthworm run might provide the means for the first seepage. With that tiny start a body of water can begin to undercut, bypass and wave a gurgling goodby to the most imposing dam.

Experienced engineers console us with the assurance that there is no such thing as a leakless dam. We are inclined to agree. Scott began damming up a small brook in Morris Run, Pennsylvania, when he was about five or six years old. He has been building dams at one place or another ever since. We have studied beaver dams and visited such notable works as the Hoover Dam in the western United States and the Aswan High Dam in Egypt. Everywhere the story is the same. No dam ever reaches the level of perfection that is mapped out by its planners and builders. If you fence in water and wait long enough, the confined water will find a way over, under, through, or around, and go rushing, bubbling, leaping from high to lower levels.

From its beginnings as an unkempt swamp, the pond area had provided a home for bullfrogs and peepers who began, while ice was still melting in the spring, to enliven the neighborhood with their peepings, croakings and serenades. They still sound out every spring. Backwater is a logical place for frogs' eggs and for the generations of tadpoles that come with each spring. Frogs like shallow water and mud. Much of our pond is less than two feet deep—a real frog paradise. They move away temporarily when we drain the pond once a year, and cut and pile up the sods that have accumulated in the bottom. When the sods have dried a bit, we wheel them to the sod-pile section of our topsoil reserve and use the resulting rotted sod for gardening. Each freshet deposits its quota of sand and silt in the pond. Each midsummer finds us stripping a quota of pond bottom and piling up topsoil for compost making and other purposes.

The Chinese drain their ponds and ditches, making special provision for the survival of their fish. They clear ditches and ponds of all accumulated refuse. When the ponds are emptied of silt, the gates are put back in the dams and the ponds are refilled. The accumulated silt and usable trash from the bottom are spread on the land or on compost piles. The product, as with us, will be a pile of first-class rotted sod which can be used in greenhouse, in mulching fruit bushes and trees, in compost making or transplanting.

What we are doing on a few square yards of a North American farm, the Chinese are doing on a nationwide scale. They are planning their agriculture, dovetailing it with the changing seasons and the weather and building their farming base.

What the Chinese are doing on a national scale, the Soviet Union is attempting on a continent-wide scale. Eurasian rivers, among the largest on the planet, which have flowed north for millennia are being turned around, made to run south into the Central Asian deserts. Twenty years ago this was an engineers' dream; today millions of desert acres are being irrigated and cultivated. Another twenty years and the desert wastes of barren

Central Asia may be feeding and clothing a great section of the human race with its harvests of cereal grains and cotton.

Our tiny postage-stamp pond is a miniature reclamation project that offers us exercise in its construction, irrigation, sods, topsoil, ice to skate on in winter and a major capital asset in case of fire. We began our work on the pond in 1953. Twenty-five years later we are still excavating, deepening, enlarging.

photo: Richard Garrett

"There are some People that care for none of these Things, that will enter into no new Scheme, nor take up any other Business than what they have been enured to, unless you can promise Mountains of Gold."

Jared Eliot, Essays upon Field-Husbandry in New-England, *1760*

"To have a great capital is not so necessary as to know how to manage a small one, and never to be without a little. It is not large funds that are wanted, but a constant supply, like a small stream that never dies."

William Cooper, A Guide in the Wilderness, *1810*

"Gardening is not only an innocent and healthy, but a profitable occupation. It is not alone by the money which is made, *but also by the money which is* saved, *that the profits of a pursuit should be estimated."*

Thomas G. Fessenden, The New American Gardener, *1828*

"A certain amount of money, varying with the number and empire of our desires, is a true necessary to each one of us in the present order of society; but beyond that amount, money is a commodity to be bought or not to be bought, a luxury in which we may either indulge or stint ourselves, like any other. And there are many luxuries that we may legitimately prefer to it, such as a grateful conscience, a country life, or the woman of our inclination."

Robert Louis Stevenson, Men and Books, *1888*

"Do not give up in despair because you have a small income and resign yourself to living meanly, in a hand to mouth fashion. Self denial and saving and resolute abstention from luxuries will solve the problem."

Mary Hinman Abel, Practical Sanitary & Economic Cooking Adapted to Persons of Moderate & Small Means, *1890*

"Whereas it matters little on Medlock Farm whether the cost of living goes up or down—it is not so much the market price of a dozen ears of corn that concerns us as that we have our own corn on the cob. No matter how low it goes it will still be cheaper to grow it than to buy it."

Henry Tetlow, We Farm for a Hobby and Make It Pay, *1938*

Chapter 8

OUR CASH CROP: BLUEBERRIES

HOMESTEADERS in the United States, as elsewhere, need a cash crop. Scrimp and manage as they will, they cannot live in the midst of a money economy without using some cash money, if only for the purchase of postage stamps.

We produce 85 per cent of our food and all of our fuel, except gasoline for the car. We must pay cash for spare parts, replacements, hardware. We pay our rent when we pay our local taxes. Some of our clothes we make, some we buy in thrift shops and at rummage sales; a few clothes we buy new. We use and buy no habit-forming drugs, including alcohol, tobacco and caffeine. Our supply of printed matter, postage and stationery comes to us via our Social Science Institute, to which organization we hand over all royalties and lecture fees. Our travel expenses are paid by those who ask us to talk.

Surrounded as we are by a cash-credit economy, we need a certain amount of cash income each year. If the amount of needed cash can be figured out in advance, we can stick to our rule of no credit purchasing and no interest slavery.

We went to Vermont expecting that our cash income would come from our woodlot: saw logs, firewood, poles and posts, greens for decoration, pulp wood. Our first year in Vermont convinced us that the easiest way to provide cash was to make maple syrup and convert a good part of our syrup crop into maple sugar. This we did for years.

Our Maine farm does not have a dozen mature sugar maples on its entire acreage. After several years' experience with selling lettuce, spinach, asparagus, peas and other vegetables, we decided in favor of berries as our cash crop.

Like Maine, our area of Vermont had been largely occupied by wild blueberries and huckleberries. We frequently discussed the possibilities of blueberry culture with our Vermont neighbors. Our experimenting began in a small way with less than a hundred two-year-old hybrid blueberry plants, carefully selected for their frost hardiness. Only in the third or fourth year does a plantation of hybrid blueberries begin to pay its own way. When we left Vermont in 1952 this experimental blueberry plantation had begun to bear substantial crops. Our Vermont experience showed us that the hardier hybrid blueberry plants would survive and produce satisfactory crops even with winter temperatures of 45 degrees below zero.

Hancock County, Maine, where we had settled, was one of Maine's "blueberry counties." Many Hancock farms included extensive tracts of wild blueberries which were burned over every second year, and in the alternate year yielded good crops of wild berries, which were an important local cash crop. The local county agricultural agent each year sent out a series of letters to growers of wild blueberries giving advice, particularly about spraying and dusting.

One of our first moves in Maine was to visit the local county agent and consult with him about the advisability of hybrid blueberries as a cash crop in Maine. His advice was brief and decisive: "Don't waste money or time planting them; they won't survive our cold winters." As we had grown them in the much colder Vermont climate, we thought they would survive in Maine, so we started with a few plants in our garden and then chose a quarter-acre plot of sandy loam, sloping to the south and west, lying to the east of our chosen pond site. The area had not been plowed or cultivated in recent years. Some of the white birch and spruce trees on the plot were a foot in diameter.

We cleared about a hundred feet square of this vigorous young forest. We cut all trees and brush as close as possible to the ground. We carted the brush away and piled it in a hollow along a small adjoining stream. The trees we cut up for firewood. In the autumn we mulched the entire patch with a layer of spruce sawdust and piled on as much spoiled hay as we could gather together. We did not plow, harrow or otherwise turn over the land. We dug no stumps. We merely planted around them. (In the early years the blueberry land was full of tree roots. As time passed and we continued mulching and weeding, the roots rotted out, enriching the land. After a dozen years the patch was virtually free of stumps and roots.)

In the spring we set stakes 6 x 6 feet each way, dug good-size holes, and set in 228 two-year blueberry plants, filling in the topsoil around each plant and tramping it hard. We used no fertilizer.

There are fifty or more named varieties of hybrid blueberries, ranging from early to late, from small to large, from hardy to delicate. We set out twelve varieties, with some from each group. By so doing, we extended our picking season from late July to the freezing point in late September or early October.

How many hybrid blueberries do we pick each year? In 1957 we picked 5½ quarts; in 1958, 60 quarts; in 1960, 120 quarts. In a word, it was seven years before we had blueberries for sale. Thereafter the pick rose steadily to 655 quarts in 1965; 1034 quarts in 1970; and 1296 in 1971, which was our banner year. Since then we pick around 800 quarts.

Our blueberry bushes are petted and pampered. The bushes are well pruned and well fed. At the moment, our earlier planted bushes are 6 x 6 feet apart; the later planted bushes, when we found out how large they grew, are 7 x 7. We would recommend 8 x 8 feet apart, as our larger bushes are already crowded and must be heavily pruned to fit into 6-foot rows.

Like most fruit bushes or trees, hybrid blueberries once set are there for at least twenty to twenty-five years. After the second year they cannot be moved to advantage. They may be dug out and replaced. But if new two-year plants are set in their places, it will be at least four or five years before a crop of any size can be harvested.

We hold the height of the bushes to seven feet, which is the height of the cedar posts which support wires and the secondhand nylon fishnet that excludes birds during the picking season.

Ideally we allow each bush to consist of six to seven major trunks, on the side branches of which the fruit buds and berries grow. Of these six to seven laterals, some should be new wood from the previous year and three or four can be older stems. This method allows for a complete replacement of the canes every third year. Since the best crops of the best berries are likely to be borne on second-year wood, this method of continuous bushtop renewal gives us an advantage parallel to that of the grower of wild berries who burns off his tops every second year.

While we are on the subject of pruning, we might mention that the hybrid plants in our area are often the victims of a parasite called witches'-broom, which grows in alternate years on fir trees and blueberry plants. Witches'-broom grows sometimes from the root; at other times it gets a start on a branch high on the bush. Wherever it appears, we cut out the soft brown twigs.

Early in our experience with blueberries we fertilized at pruning time in the spring and again at the end of the crop season. But new growth, stimulated by this procedure at the end of the growing season, is likely to be frozen out during our severe winters. Latterly we have fertilized only in the spring. Our pruning is done during the good days in late winter or in the spring before the buds have begun to swell. We prune and fertilize heavily in order to increase the size of the berries and the volume of the crop. Under these conditions we get a reasonably abundant crop each year.

Spring feeding follows spring pruning. It consists of about one pound per bush of a meal composed of soybean meal if we can get it. When soy meal is scarce or excessive in price, we use cottonseed meal, linseed meal and as a last resort cornmeal. With the soy we mix equal parts of ground phosphate rock and granite dust. Over the meal we spread compost, eight to fifteen pounds per bush, depending on size and appearance. We top off with sawdust (about a peck for a good-size bush) and then spread hay or straw in the spaces between the bushes. As the hay is tramped down by weeding or picking we replace it.

Blueberry bushes prefer a sandy loam or gravel soil somewhat acid in character. We use no lime and no wood ashes on our blueberry plants and aim to keep the soil on the acid side. Each feeding includes a generous ration of peat moss per bush. In Michigan, where peat bogs abound, we have seen an entire blueberry plantation in a peat moss bog. The plants were about ten years old and seemed to be doing well.

Who picks our berries? Generally we pick them ourselves, as we know which are riper; if the berries are picked too early, they are sour. Sometimes the berries are picked on shares, one quart out of four for the picker. Pickers who are newcomers generally go for the biggest berries and leave the smaller-fruited bushes. They also eat a lot of berries. One woman who came to buy berries asked, "If I pick them myself, can't I have them for less?" "No," said Helen. "The price should be higher because you'll eat so many."

Picking is easy and pleasant. With the birds singing and the frogs croaking and the wind blowing and the sun shining, it is nice work. With good picking conditions (berries ripe and abundant and not too much conversation), one can get ten or twelve quarts in an hour, but not everyone can move that fast.

Blueberries make us a moderate cash crop. They ripen on the bushes and (unlike raspberries, which must be picked as soon as they are ripe) they stay there until they are picked. Taken into our cold cellar, they remain in good condition for days or even a week if necessary, without apparent disadvantage or loss of firmness or flavor, which cannot be said of strawberries or raspberries.

Everybody likes ripe blueberries, even our native neighbor friends who first scorned them as "not up to the wild ones; tasteless." After a few presentation quarts they now come and buy them. Animals like blueberries too, especially deer and raccoons and a great variety of birds, including the seagull—a big bird with a huge appetite.

Since our blueberry plants began bearing, we have never had a crop failure. The pick has been greater in some years and smaller in other, depending largely on the weather. But each year there is a sizable and marketable cash crop.

"I am the warmth of the hearth on cold winter nights. I am the shade screening you from the summer sun. My fruits and restoring drinks quench your thirst as you journey onward. I am the beam that holds your house; the door of your homestead; the bed on which you lie, and the timber that builds your boat. I am the handle of your hoe, the wood of your cradle, and the shell of your coffin."

Sign on a tree in a public park in Madrid, Spain

"A tree: the grandest and most beautiful of all the productions of the earth."

William Gilpin, Remarks on Forest Scenery, *1791*

"The duty of a forester consists in preserving order and beauty, furnishing timber or copse, and providing a succession of young trees for falls of timber, additional plantations, other uses, or decay or accident in any part under his charge."

John Loudon, A Treatise on Forming, Improving & Managing Country Residences, *1806*

"There are few farms in the United States where it is not convenient and profitable to have one or more wood lots attached. They supply the owner with his fuel, which he can prepare at his leisure; they furnish him with timber for buildings, rails, posts and for his occasional demands for implements.... In most woodlands nature is left to assert her own unaided preferences, growing what and how she pleases, and it must be confessed she is seldom at variance with the owner's interest."

R. L. Allen, The American Farm Book, *1849*

"According to the common estimate of farmers, the woodlot yields its gentle rent of six percent, without any care or thought, when the owner sleeps or travels, and it is subject to no enemy but fire."

Ralph Waldo Emerson, Country Life, *in* Natural History of Intellect (vol. XV), *1904*

"So far we have concentrated most of our bucolic attention on food: That is as it should be because it is the most important thing the land produces. But it also yields a number of non-edible things: the woodlot, for example. Year for year, the experts say, trees pay out better than grain crops."

Henry Tetlow, We Farm for a Hobby and Make It Pay, *1938*

CHAPTER 9

TREE CROPS IN MAINE

TREE CROPS have played a significant role in the life of human beings for a great part of written history. Evidently, trees will continue to be one of the basic resources available to mankind.

Before men learned to domesticate animals or to cultivate the earth, they secured an important part of their food supply directly from trees. They used tree leaves, tree roots, tree sap, tree bark, tree shoots, tree flowers and tree fruits for food. As the use of fire became widespread, wood, which is derived from the bodies of trees, was used for warmth and for food conditioning. Until recently, tree bodies in the form of lumber were the chief structural material. Even today, in an era dominated by minerals and metals, wood has a great variety of uses.

Every climatic belt inside the polar extreme has tree crops for which it is peculiarly suited. Maine is no exception. Apple trees grow wild all over Maine. Originally they were crab apples, with a minimum element of usefulness. Johnny Appleseed and his contemporaries introduced into New England fruit trees and berry bushes developed in Europe, elaborated in North America and presently providing the varieties that dot the Maine countryside. In the hedges and thickets, along the highways, in former pasture lands presently growing up to brush, there are endless volunteer apple trees.

Our farm in Maine is well situated for growing hardy fruit. A century ago a large part of the farm was occupied by a prosperous apple orchard. When we bought the place in 1951 a few of the original trees from this old orchard were still bearing. They were splendid specimens of orchard antiques—some of them must have been at least a hundred years old.

Where we have cleared on our Maine farm we have found scrub apple trees among the brush, holding their own in the growing forest on our woodlot. Several of the volunteer apple trees stand out each spring crowned with masses of apple blossoms.

When we do our annual clearing in the woodlot, if we find promising apple trees we clear an open space around them, prune the trees as we would any promising young fruit tree, thin out most of the fruit buds and check on the remaining buds to see if they will fill out. We have observed that when apple trees grow from whips into promising young trees, they behave like any forest tree. They grow up with the neighboring shrubs and trees, competing for a place in the sun.

One apple tree near our blueberry patch was of this forest type, with blossoms and fruit at the top, forty feet above the earth's surface. The tree at stump level was about eight inches in diameter. The tree looked like a vigorous grower, so we cleared around it and left it standing by itself when we put in our blueberry plantation in 1953-54. The tree bore a Greening-type apple, of moderate size. The crop was borne on the top of the tree. Almost every apple was marred by destructive apple scab.

Could we convert this wild thing reared in a young forest into a disciplined and productive bearer of the kind of fruit that appeals to human beings? We decided to experiment.

We climbed to a point on the apple tree about fifteen feet from the ground where there was a whorl of side branches and sawed off the top of the tree at a point where it was about five inches in diameter. In a word, we cut the tree in half and took off the entire top, which had been bearing most of the blossoms and fruit. We pruned the lower half of the tree into some semblance of an orchard tree. For three or four years the tree bore few blossoms and little fruit. Today the Greening tree looks like an orchard native and is bearing a good crop of good apples.

We found a pair of what we consider volunteer apple trees standing rather close together when we took over the Maine farm. The trees were near our garden site and close to our compost-making area. Both trees were so close to a ditch that their roots were being undercut by spring runoff. One of them was leaning at an angle of 20 degrees. Both trees were volunteers. Both bore Cortland-type fruit, but so far as we know they had not been grafted. The leaning tree was the more vigorous grower and was bearing larger crops. Apples from both trees make good juice and applesauce.

We began with these trees by moving the stream well away from their roots. We fertilized and mulched them generously. Size and quality of the fruit increased. Apple scab, which had been bad in both trees, diminished notably, although we did not dust or spray. We thinned the fruit rigorously, picking off scabby and wormy apples. In good apple years, they produced large crops. In the 1975 season, a rather good apple year for us, we picked twenty-five bushels of apples from the two trees. Size of apples was good, and the quality was excellent. They kept in the root cellar until early spring.

Thus far we have been writing as though the chief product of forestry were the fruit yielded by trees, when as a matter of history and experience the chief crop yielded by forests is wood: wood for fuel, wood for building, wood for furniture, wood for paper, wood for a thousand and one objects.

There was a time not many years ago when important parts

of the Americas, Asia, and Africa were still occupied by virgin forests grown through centuries. Today all of the remaining forested areas are under heavy pressure to turn standing timber into a cash crop. As the world's population increases, the need for wood increases.

Large areas of the earth's land surface are better adapted to producing tree crops than any other product. Steep hills, narrow valleys, thin soil, rough climatic conditions make farming and gardening difficult. For such areas tree crops are natural. Much of New England falls into this category. For ages it has produced high-quality forest products. If the human beings moved out of New England for a century or two, we have every reason to assume that similar forests would again establish themselves.

The Chinese and the Albanians in their overall thinking about land use have developed this formula: on mountains and steep slopes up to the treeline, forests; on more gradual slopes, orchards and vineyards; on land that is flat or nearly flat, open cultivation. Following such a formula, much of Maine would be forested; its more gradual slopes would bear orchards and vineyards. Only relatively level land would be subjected to open cultivation, without the consequent threat of erosion.

Homesteaders who settle in New England and who take for granted the need for tree crops will see to it that the weeding and thinning of the woodlot occupies a high place on the list of priorities.

In trimming out our woodlot we follow a simple formula. We decide what trees will do best in a given area—swampy, rocky, hilly. How close together should trees stand in a young, growing forest? The answer depends on the age of the forest and the type of tree. Balsam for the Christmas-tree trade can be planted or thinned to 6 x 6-foot distances. When of marketable size, every second tree can be taken out. Trees should not crowd each other, but the sun can be almost excluded from the forest floor. In the garden we refer to "thinning" and "weeding." In the woodlot the same principles and the same terms may be used.

In a moderate-size Maine woodlot, our chief objective is to provide logs for milling, fuel for heating and cooking. Wood for heating should if possible consist of hardwood. In our region near the coast, hardwood is scarce. We have no oak or beech—two of the trees that provide wood with a maximum of heat units. Black cherry and white ash come next. Then come spruce, hemlock and balsam fir.

Felling trees is a simple matter if axes and saws are in good order. Small trees, up to eight or ten inches in diameter, present no problem. Larger trees, particularly if there is a thick stand in the woodlot, may be troublesome and, in the case of very large trees, dangerous. A competent woods worker can usually determine almost exactly where a given tree will fall when it is properly notched and sawed.

Once the trees are marked and down on the ground, all branches are chopped or sawed off. The trimmed tree trunks are sawed into lengths. If the project is a commercial one where high wages are paid, tree trunks are cut into logs and pulp wood. In our small-scale woodlot operation, economy is the determining principle. We cut some logs and pulpwood, but mostly firewood for heating and cooking. So we save branches that a commercial lumberman would scorn, and cut them into stovewood lengths. For us, good limb wood, well dried, is as good firewood as you will get anywhere. It cuts up easily and burns magnificently. For hot fires we know nothing better than well-dried limbs.

About four-fifths of our Maine farm constitutes a woodlot that matures year by year. About one-fifth is cleared land that we mow or scythe each year to keep back the encroaching woods and maintain the grasslands in hay for mulching and compost making.

We do not own a mowing machine (which we would use perhaps six or eight hours in the year), so we try to hire a neighbor to come in with his machine for an hour or two. As there are few functioning mowers in the neighborhood, we have equipped ourselves with scythes which we use in the most necessary places. Early every morning for about two months Scott is out swinging the scythe for his pre-breakfast exercise. One notable summer day last year, eight or ten neighbors contributed their time and energy and skill to hand-scythe from dawn to dusk and cleared a whole meadow for us. It was a Breughel-like scene with figures of men, women and children moving over the field.

In cutting trees or keeping back encroachments on meadows, we have enlarged the cleared area of our farm somewhat during the past quarter-century. We regret each acre that we pilfer from the reforesting process. Nature, here in Maine, moves from the bare earth to splendid tall timber. We regret any steps that we take in opposition to that process.

Almost everyone who settles in New England is tempted to clear some land. Trees, big and small, perennials and annuals, are cut off, their roots are destroyed and the earth is left bare and open to elements which immediately set in motion the forces of erosion. Wind blows; water runs; sun burns; fire destroys. Under such conditions one would think that in the course of time the soil would disappear and only the rocky skeleton of the countryside would survive. Land clearing, plus brush and trash burning, should have brought rocks to the surface years ago. But that just does not happen even on waterless land. On the contrary, the hills and valleys of New England have attracted and sheltered a multitude of plant life and wild animals for ages.

Instead of erosion seizing the initiative and disposing of every thing movable or burnable, natural forces make soil, enrich it and utilize it as a base for: the first year, annuals; the second year, shrubs; in a decade, young saplings on the way to becoming stately trees.

Anyone who is rash enough to try to keep New England free of trees and foliage has taken on a Herculean task. You may interfere with the process, but you cannot stop it. You might as well try to cure water of the urge to run downhill. Under existing conditions you cannot keep an acre of Maine land free of vegetation without cutting, clearing and mowing. If you let a back pasture lot go unmowed and untended for a time, it passes through a regular sequence from grass through brambles to brush composed of alder, birch and minor softwood, to semi-hardwoods like ash and soft maple and then, if the soil is ready, to yellow birch, and hard maple and ultimately to spruce, pine, hemlock and other softwoods.

The forces which made the greening-up process unavoidable, operating through centuries before Europeans came to North America, had covered the state of Maine with a blanket of green that extended from ground cover to magnificent tall timber. There was a time, not too long ago, when some of the finest ship-building timber on the planet was growing in Maine. At that time, Bangor, Maine, dubbed itself lumber capital of the world, basing its claim on the boast that it handled more lumber than any other city in the world. Maine's virgin forests have been devoured by the increasing demand for wood and still more wood. Today, in Maine, mature trees are a rarity. The murmuring pines and the hemlocks which once covered the bulk of the state exist only in song and story. The sequences of forest reproduction are interrupted as soon as young trees have reached a size that will make pulp wood. They are rushed to the pulp mills and converted into newsprint that announces price-cuts on supermarket bargains. For first-class lumber Maine now imports from the southeast, the northwest, Norway, Sweden and the Soviet Union.

If human beings would keep their hands, axes and saws to themselves for a very few generations, the Maine climate and soil, left to the normal processes of forest reproduction, would once again reestablish a magnificent growth of superb timber.

photo: Richard Garrett

"The greatest value is received before the wood is teamed home.... It warms us twice, and the first warmth is the most wholesome and memorable, compared with which the other is mere coke."

Henry David Thoreau, Journal, *October 22, 1853*

"I found chopping, in the summer months, very laborious. I should have underbrushed my fallow in the fall, before the leaves fell, and chopped the large timber during the winter months, when I should have had the warm weather for logging and burning, which should be completed by the first day of September. So, for want of experience, it was all up-hill work with me."

Samuel Strickland, Twenty-seven Years in Canada West, *1853*

"There is a good deal of difference between a timber forest and an ideal farm woodlot. The latter must serve a variety of needs. The more different kinds of wood it contains, the better; and there should be trees of all sizes and ages, so that the production of available material for various uses will be steady. Our biggest single need is for firewood. Although we use four fireplaces pretty steadily from early fall into later spring we have seldom had to cut up a whole mature tree for fuel alone: weed trees, fallen trees, and the topwood of trees cut for posts or other purposes supply ninety percent of it."

Henry Tetlow, We Farm for a Hobby and Make It Pay, *1938*

CHAPTER 10

WOOD FOR FUEL

WOODING-IT (in the vernacular) has been a basic source of income, a pleasant leisure occupation and a health-preserving avocation ever since human beings learned to make and use fire for their purposes. The practice is still popular, even in centers dedicated to technology, mechanism and automation. Wooding-it heats homes and helps in food preparation every day in millions of homes and homesteads all over the planet.

Wooding-it can be practiced by any homesteader or householder. It requires only a few simple hand tools which anyone can learn to use effectively. It occupies spare time that might otherwise be spent in front of the TV and keeps even decrepit oldsters out in the open while they make a real and substantial contribution to family income and to family comfort around the crackling open fire.

On a recent visit to England we dropped in to see a friend, Gordon Latto, who has a home in the country within commuting distance of London. No sooner had we finished our exchange of greetings than, at Scott's request, Gordon took him to a corner of his side yard and introduced his woodpile. A neighbor had been making some housing alterations that necessitated the cutting of several trees. Instead of letting the trunk and large branches of these trees go to the dump, Gordon had the good sense to have the drayman bring them to his backyard. There they lay, ready to be sawed to fireplace length. When cut, piled and dried out, the wood could be wheeled to a woodshed which opened directly on to the living room with its open fireplace.

Scott rolled up his sleeves, thanked his host for putting two newly filed saws and a sharp axe at his disposal and spent the rest of the morning happily fitting the wood to the fireplace. After days spent in airplanes, and nights in hotel rooms, the relief to him of sawing wood for hours was truly wonderful.

Horace Greeley, onetime editor of the *New York Tribune,* loved wooding-it as much as Scott. In 1868, he wrote in his *Recollections of a Busy Life:* "The woods are my special department. Whenever I can save a Saturday for the farm, I try to give a good part of it to my patch of forest. The axe is the healthiest implement that man ever handled, and is especially so for habitual writers and other sedentary workers, whose shoulders it throws back, expanding their chests, and opening their lungs. If every youth and man, from fifteen to fifty years old, could wield an axe two hours per day, dyspepsia would vanish from the earth, and rheumatism become decidedly scarce. I am a poor chopper, yet the axe is my doctor and delight. Its use gives the mind just enough occupation to prevent its falling into revery or absorbing trains of thought, while every muscle in the body receives sufficient, yet not exhausting, exercise. I wish all our boys would learn to love the axe."

Wooding-it has another advantage of overriding significance. From the decision to cut or not to cut a certain tree, through all the processes of felling and fitting, wooding presents a series of choices, decisions, tests and experiments that often preoccupy or perplex the woodworker. With experience, shrewd judgment, quickness of eye and nimbleness of foot, the successful woodsman, in half an hour or half a day, can point to a heap or pile of fitted wood as a concrete testimonial to his competence and persistence. Any tree correctly felled, any log properly split and stacked, gives a chance to evaluate, test out and get results time after time, hour after hour. Each tree, log or stick is a problem in its own right with its possible outcome of usefulness. In the north temperate zone, wooding-it will continue to be an important source of livelihood for a long time.

We cook and heat with wood. Since we have a woodlot, the outside energy needed to keep us fed and warmed in cold weather comes from a source under our own control. Wood and stoves make us independent of imported fuel.

Where we live in Maine we estimate our woodlot needs thus: a family-size cookstove will consume about four cords of moderately good dry wood in the course of a year; a chunk stove able to heat a couple of rooms in a house will require an equal amount of moderately good wood in a year.

Well managed, a growing woodlot of ten acres properly thinned and weeded should furnish an adequate supply of wood if the fuel is stored and well dried. We have two chief sources of wood. The first is our woodlot of about fifteen acres of young and old growth. The second is the driftwood that each tide and especially each storm brings into our cove. Occasionally, whole trees drift in.

Last year one of the logs that ended up in our woodshed was fifty-one feet long and two feet in diameter at the large end. It was water-soaked and consequently very heavy. We guessed that it was hemlock because of the large number of enormous knots. We sawed it into twelve pieces on the beach, loaded them with difficulty into our pickup truck and hauled them to our wood yard, where they stood around like huge Stonehenge blocks for a season while they dried out. They were then split and stored away under cover.

Usually tidewater wood comes in the form of logs or four-foot billets cut for firewood or for the paper mills. Many used timbers from demolished wharves and jetties and the construction industry end up on our beach. About half of the wood we burn is driftwood. Even when dried it is not good wood for stoves or fireplaces, since it produces creosote which clogs the chimney. But we enjoy using up an otherwise waste product. It also cleans up the beach.

Products of the woodlot and of tidewater are cut to three lengths: logs, which are either 8 or 12 feet long, billets, which are either 38 or 48 inches long; and miscellaneous wood, which is cut into stove lengths.

Our wood needs are governed by the size of the fireboxes in our heating and cooking stoves. Our heaters can handle wood up to 28 inches in length and our kitchen stove firebox will take nothing longer than 16 inches. To economize time and facilitate handling our wood supply, we cut all of our wood to a uniform length—16 inches for kitchen wood and 19 inches for chunk burners. The 16 inches is determined by the firebox. The 19 inches is the result of a decision to let wood go to the woodshed in only two lengths—kitchen and chunk burner. If all wood comes in these dimensions, piling is relatively easy. If wood comes in random lengths, piling is more laborious and unsatisfactory.

Conditions for drying wood are not too favorable in coastal New England, with its fogs, mists, high overcasts, fierce winds and nagging precipitation. For a brief period after wood is cut, it may be left in the open to season, harden and, if there is any sunshine, dry off superficially. For best results, both cooking wood and wood for heating should be kept under cover.

In the drying-off stage, we cut tree trunks and branches 5 inches or more in diameter into 8-foot lengths and pile these on skidways. A skidway consists of two poles or small logs laid 4 feet apart and blocked up sufficiently to keep the bottom logs from being rotted by contact with the earth. If there are many logs, it pays to divide them with stickers, poles or odd bits of lumber so that there is air space between the lower and upper layers of logs.

As a result of our experience in the mountains of Austria, instead of laying limb wood and poles on the ground where they attract moisture, we stand them on their butt ends in teepeelike formation. Logs and billets we cut to size. Everything else, long or short, goes into a teepee if it is 4 inches or less in diameter at the butt end.

Teepee construction is simple. We drive an 8- or 10-foot stake into the ground. Against this stake we lean poles and limbs, butt down, in concentric circles. Larger teepees are 8 or 10 feet in diameter at ground level. Some teepee poles are 18 or 20 feet long. Teepee wood, instead of lying on or near damp ground, stands up in sunshine and fresh air. Rain and moisture run off it and it dries out quickly.

Since we live in New England weather conditions, which make dry wood almost a must, in setting up a homestead we therefore set aside part of a building or a separate building for wood.

The woodshed itself is of the simplest construction, open on four sides to let in a maximum of fresh air and sunshine. Our main woodshed is built on a rock outcrop. We picked four outcropping rocks, cleared them carefully and built on each of the rocks a wooden form. The four rocks and their piers made a rectangle 12 x 14 feet. The four forms were built on the uneven rock outcrops so that the tops were at the same level (10 to 18 inches from the ground). Into these forms we poured concrete made of six parts gravel, three parts sand and one part cement, well mixed into a sticky mass.

At the center of each of the four piers we set a piece of iron one inch in diameter and about two feet long—half in the form, the other half standing vertically in the air—at each of the four corners of the prospective woodshed.

We planned to have the woodshed about 7 feet above the

ground at the plates. So we cut four cedar logs 7 feet 6 inches in length, bored a hole 12 inches deep in the top end of each log and, as soon as the concrete piers had hardened, set the four logs, butt end up, on the inch bolts embedded in the concrete.

We braced the four corner posts into position. We divided the top of each post into four segments and cut away the three outside segments, leaving the inside quarter of the post with a 6-inch projection and two shoulders to carry the 6 x 6-inch spruce plates. These plates had been hand-hewed at the ends, ready to make the halved corner joint at each of the four corners.

One-inch dowel pins were driven vertically into the corner posts and horizontally into the projecting inside quadrant of each corner post. Each corner post was braced into position by two 6 x 6 hand-hewed spruce timbers, their ends cut at 45 degrees, and pinned into position by one-inch dowel pins.

The ridgepole was a spruce hand-hewed timber set on the plate at each end and held up in position by a 6 x 6-inch hand-hewed post.

The next operation was to drive a ring of twenty-penny nails halfway into each corner post, lift the corner forms four inches and pour a stiff gravel-sand-cement concrete into the elevated forms, tying the four piers, the four posts and the four plates firmly to the rock outcrops on which the new structure was to rest.

As good luck would have it, a building contractor was tearing down an old and very well built house in Harborside. Into one pile he had heaped the 2 x 6 x 12-and 2 x 6 x 14-foot roof rafters, which bristled with fragments of broken wood and projecting rusty nails. It was as unsightly a mess as one could imagine.

We stopped by to have a word with the contractor and noticed the heap of old rafters. Said the contractor: "Anyone with a twenty-dollar bill can have that heap of trash." We produced a twenty-dollar bill, loaded the nail-studded rafters into our truck, drove home with our treasure trove, pulled out a thousand and one nails and had enough sound 2 x 6 timbers to provide rafters for the new woodshed.

The building was to have a roof of 28-inch by 12-foot aluminum sheets. The load of 2 x 6 scrap rafters from Harborside not only gave us rafters for the new building but gave us enough sound lumber to provide 2 x 6 nailing strips to carry the aluminum roof. By using the 12-foot rafters we were able to provide an overhang of almost 4 feet on the two drip sides of the woodshed roof.

We therefore had a woodspace of 12 x 14 feet, plus a comfortable, relatively dry space under the overhanging eaves on both drip sides. In this dry space we could lay out wood and work it up on rainy days, or stack wood temporarily while it was awaiting piling space.

The four sides of this woodshed were four piles of two-foot wood, laid ends to the weather. Inside this covered, woodpiled shelter, through which the wind whistled but into which only the finest snow could be driven by a high wind, we had a workplace for rainy days that was gradually filled by piled wood as we accumulated our supply for the succeeding winter.

In the woodshed, in addition to an old axe and a pulp hook, we keep a set of measure sticks about the size of ordinary wooden rulers. Each is plainly marked with a crayon—16 inches for the kitchen firebox, 19 inches for the chunk burners, 24 inches for the schoolhouse stove. The measures have holes bored through them and hang together on a nail driven into one of the corner posts.

With a measure stick in the left hand and a 30-inch bow saw in the right hand, we are prepared for anything up to a log or billet about 8 or 9 inches in diameter.

If a half hour comes along on a rainy day with nothing else to do, we can spend time in the woodshed under cover and with something practical and productive to show for each saw stroke. In addition to three 30-inch Swedish pulp saws, we have a 36-inch and a 42-inch Swedish bow saw which we use for logs more than 8 inches in diameter.

One of the most difficult jobs associated with self-sufficiency on the land is keeping the place and the tools neat and tidy and in place. On a general farm, even a small one, this is a constant problem. Each season has its particular needs, and each job has its appropriate tools. Sometimes, even from day to day and hour to hour different implements are needed and should be in place.

In the woodshed, in addition to axes and pulp hooks, we keep saws, shovels, forks. We keep our saws sharp and have a place to hang each saw when it is not being used. If tools are always cleaned after use and put into place, they are likely to be found when needed. Kept sharp and in good order, these tools can last a lifetime. In our woodshed we still have a double-bitted axe with a 30-inch hickory handle that Scott used as a young man around 1900. To be sure, the axe is worn down, stumpy and now used chiefly for grubbing, but the handle is still in one piece and it's a handy tool to have around the place.

After every morning or afternoon of work we gather up the tools we have been using, clean them with strips of burlap cut for the purpose and put them into their allotted place in the woodshed. A big machine like a cement mixer needs washing and scraping after every large or small job if it is to be kept in condition. We had one that turned by hand and mixed concrete faster than we can mix with a shovel in a wheelbarrow, but it was too much trouble to keep clean and to house. When we found it standing idle after some years, taking up room in the shed, we gave it away.

Wheelbarrows are among our favorite tools. We maintain a battery of four metal contractors' barrows with head-on dumpers attached. We use them generally for moving bulky or heavy materials. For all our concrete work in Maine we used wheelbarrows. Only during the last few years did we discover a new plywood barrow made in Vermont. It has two wheels and therefore takes a wider runway than a single-wheel wheelbarrow and carries more of a load. For light, bulky material like hay it is vastly superior to a wheelbarrow, although it takes up more space when stored.

If you occupy a cabin on a ship, or the equivalent in space, you have small choice but to keep things tidy. Space is limited and superfluous objects simply cannot be tolerated. But if you live on one or more acres, there are immeasurable possibilities for litter and clutter. With a front yard, a back pasture and a woodlot to fill up, stray pieces of wood lumber, tin cans for the dump, an odd bit of worn-out machinery, a car minus some tires, a broken cultivator or hay rake appear here and there until the whole place takes on the appearance of an unorganized junkyard. A modern homestead fills up with superfluity and waste that soon lead to disarray and chaos. Order in the woodlot, the woodshed, toolshed, yard and home are essential in the practice of the good life.

Unless wood is very expertly piled, it is likely to settle one way or another and, sooner or later, tip over. A pile of standard cord wood, 4 feet long, will generally stand up if it is less than 6 feet high. Shorter lengths always present problems. All wood in our woodshed is less than 48 inches in length. Our 16-inch wood is always a problem in any pile over 3 or 4 feet high. With a tight roof over our heads and some form of protecting siding, it is always possible to toss 16-inch wood at random into a stack. At the end of three or four months of such stacking, almost any wood will be dry enough to burn. In the woodsheds of both of Scott's grandfathers, this was their custom. They had the unused space—more than they needed—so they just sawed, split and tossed the firewood into stacks that reached the woodshed ceiling.

We enjoy both sawing and splitting and we are especially

happy with neat, even, erect, self-respecting woodpiles. With 4-foot wood this is no problem. With 16-inch wood it is a problem that must be faced with piles of more than 4 feet in height.

As we saw our limb wood and pole wood, we pick out some straight pieces an inch in diameter, cut them to 32 inches and have a small pile of these pieces handy as we pile. Instead of piling one pile at a time, we pile two, side by side. As the piles go up side by side, for each 12 inches of height we lay a 32-inch binder across both piles, and a binder every 3 or 4 feet sideways, and go right ahead piling over the binders. This gives us in effect a 32-inch-wide pile self-tied together as it goes up. The chances are that two such piles will stand up until needed for the stove.

When the pile of 16-inch wood has reached the required height, if it has been skillfully made it will be a work of art. If the piler wishes to take one more step in refining his artistry, he can use a piece of 6-inch board 3 or 4 feet long, stand it vertically or lay it horizontally along the completed pile and tap it with a light axe, bringing all of the ends to a line as even as any wall.

We go over all brush carefully and trim branches down to a diameter of one inch. And the branches of less than an inch in diameter? If we lived in Western or Central Europe we would treat this brush with respect. Here in New England most people leave it where it falls unless, like us, they take seriously (1) the problem of forest fires and (2) the conservation of our diminishing forest wood supply.

We are developing a technique of faggot making that follows the European tradition. Most West Europeans are hard put to get any burnable material—be it fossil fuel or wood currently produced. Consequently, like people in all parts of the world where wood is scarce, they pick up every available fragment of burnable material, including brush, and use it for cooking or heating. This has resulted in the European faggot: a bundle of twigs or branches, cut to size and bound together by twisted grasses, twigs or other likely material including waste string.

Our faggots are of two kinds: those of kindling size, used to start fires; and faggots made up of pieces of brush wood and limb wood from an inch to an inch and a half in diameter. This material when gathered should be at least partially dry; it should then be broken or cut to firebox length and bundled in units that can be added to a small amount of paper and ignited. If well made and dry, the faggots burn as readily as paper and are almost surefire with the first match.

We have built convenient troughs made of a base piece and two side pieces in which to assemble faggots. The base piece is at least 8 inches wide. Nailed to the edges are two 4-inch strips. Notches are made at convenient intervals so that each faggot is bound together by two strands of binding material. Binders are slipped into the notches; the trough is filled with cut-to-length or broken pieces and the binders fastened with easy-to-untie knots of scrap string or rope. If the string is good for two or more uses, so much the better. These bundles may contain anything up to a dozen or more pieces of brush or wood.

We know no other American homesteader who goes to the trouble of making up these faggots. We have found them invaluable in using up small pieces of wood otherwise difficult to pile and therefore tossed away or left neglected.

So much for wooding practices. Like any other activity, wooding can be just a chore and a bore, or it can give the wood handler a chance to put artistry into the trade. There are several ways to perform almost any act—an efficient, workable, artistic way and a careless, indifferent, sloppy way. Care and artistry are worth the trouble. They can be a satisfaction to the practitioner and a joy to all beholders. For us, efficiency and artistry always pay off because of the satisfaction of doing a job well.

We live in an age of quick and easy heat and power supplied by kerosene, gasoline, natural gas, fuel oil, electricity. The supply of these fossil fuels is sharply limited, less and less adequate to meet the increasing demand. All are produced and sold to consumers at high prices, which will increase as the years pass. The energy squeeze in 1973-75 created a big demand for wood and was a foretaste of what must happen as the supply of fossil fuels diminishes with population growth.

The purchase and sale of these sources of energy trespass on one of our basic formulations for the good life: "serve yourself." Those who can meet the demand for wood from their own woodlots will have an invaluable source of economic stability and security.

photo: Richard Garrett

"When thou hast chosen a convenient and fit plot of ground to digge a garden in, then must thou in handsome manner castying the utter compasse of it (as eyther four square, or round or otherwyse) enclose the same round about, and besett it and fence it throughout before you go about to dresse up or sowe any thyng within the same. For Gardens must be well fenced and closed about, before there be any thynge sowen in them."

Thomas Hyll, First Garden Book, *1563*

"All your labor past and to come about an Orchard is lost, unless you fence well. For you can possess no goods that have so many enemies as an Orchard. Fence well therefore, let your plot be wholly in your own power."

William Lawsen, A New Orchard and Garden, *1676*

"I am amused to see from my window here how busily man has divided and staked off his domain. God must smile at his puny fences running hither and thither everywhere over the land."

Henry David Thoreau, Journal, *February 20, 1842*

"When we have stones to contend with, we raise them above the surface by the help of levers. By these means, stones of half a ton weight can be more easily lifted from their beds. The larger ones are generally drawn off the fields to make the foundations of fences, and those of a smaller size are used in the contruction of drains."

Samuel Strickland, Twenty-seven Years in Canada West, *1853*

Chapter 11

STONE WALLS VERSUS WIRE FENCES

We had lived in Vermont for almost twenty years, where we had fairly friendly relations with our neighbors and a consequent de-emphasis on fences or fencing. We dislike fences on principle because they shut people and creatures out. We like wild animals. One of our first experiences in Maine was watching two does and their fawns playing on our meadow. The little deer were like overgrown puppies, chasing each other back and forth over the greensward. While the young deer raced and circled, tumbled and butted each other, the mother deer browsed and watched. It was a happy family scene.

We were delighted with the fairly regular sightings, all to be seen through our house windows. There had been many deer in Vermont but they never came near the house, which was close to the road and had no extensive meadows about it. Here was a dividend on which we had not counted.

There were wild creatures aplenty around our isolated house. Except for short summers, the house we bought had been unoccupied for a few years and the wild animals had come to accept the place as deserted. By and large they were unafraid. One raccoon came to our back door and we lured it into the kitchen with a piece of bread. It prowled about the room and had to be got out with another piece of bread.

We later found varied reasons to keep the wild things out in the woods where they belonged. In Vermont we had tried without success to grow grapes. The vines survived the hard winters but the summers were too short for the grapes to ripen. We thought we would try them out in Maine. A hundred feet north of our house there was a rock outcrop that faced directly south. This outcrop seemed to offer us just the opportunity we wanted. A well-fertilized bed along the rocky edge could be enriched. A trellis could be built over the rock, across which the grapevines could trail and where they would be protected against the north winds and have the advantage of sunlight. The rock would absorb heat during the day and radiate it at night.

Accordingly, we made a grape bed at the base of the rock ledge, bought some carefully selected grapevines and looked forward to a crop of delicious grapes by the third year. The grapevines did their part. They took root and grew sturdily. By the end of the second year our cherished vines were thriving against the rock outcrop. One more year and we could anticipate a crop of grapes.

Came the third season. The grapes had lived through two winters. They budded out properly and the young leaves began to appear. Early one bright morning we chanced to look up toward out prospective vineyard. Dawn light showed us two half-grown deer standing in front of our grape trellis nibbling the vines. We got field glasses for a better look. With dismay we watched while the deer ripped off each partly grown grape leaf. Leaf by leaf, they were being daintily plucked from the young vines. When the deer had finished their morning snack there was not a green leaf or a bud remaining on the grape arbor. That was the end of our grape venture unless we were prepared to fence in the outcrop.

Between this rock outcrop and the house we planted a dozen Rugosa roses, which we found in a nearby cove growing just above high-tide mark. Rugosa rosebushes grow, at best, to a height of five to seven feet. Once established (as they were along the coast), they thicken to an all but impassable bramble patch and therefore make an excellent hedge. They are showy green in spring, rosepink or white when in flower. By mid-August the petals are replaced by a reddish yellowish rose apple or rose haw that can be as much as two inches in diameter. When mature, these rose apples contain much more vitamin C than the best citrus. To make Rugosa still more attractive, they are a decorative shrub, beginning to flower in early summer and continuing to flower and fruit until frost cuts them off in the autumn. We are loyal friends and admirers of Rugosa's foliage, flowers and fruit.

From the dozen plants we now have thirty-five hills in our planting. Each year they are heavily pruned and thinned and fertilized generously with rotted sod dressing and a home-mixed protein meal and mulched with seaweed fresh from the shore. We prune and treat the Rugosas about like raspberries. The hills are four to five feet apart in each direction. Additional young plants, of which there are many, are dug out. Each year a third of the old plants are cut away. In their places, two or three young plants are allowed to replace those that are being removed so that each hill retains from four to seven plants.

One year at thinning time in early summer, Eliot Coleman, a near neighbor, took forty of our young plants to establish a plantation of his own. Of the lot he lost only one; thirty-nine lived and provided him with the basis for his present extended planting of Rugosas, from which he in turn gives away numerous plants.

When we began our Rugosa patch, we did not fence it. By the third year the deer had discovered it. That year they sampled the rose apples. The next year they began on the tender leaves early in the spring. By the following year they were eating the leaf and flower buds from the roses. That year we did not get a single rose hip from our plantation. That same year we put up a wire fence.

One more item in this list of reasons that led us to fence our vegetables and fruit: Traditionally we plant sweet corn—usually from eighty to a hundred hills. For a number of years our corn was molested only occasionally. Finally this molestation became serious. One year, from a hundred corn hills we got not one single ear of mature corn. All were destroyed by raccoons before they ripened.

Rabbits, woodchucks, squirrels, porcupines all caused damage so serious that we had to adopt countermeasures or do without, as in the case of grapes and the sweet corn. After much discussion and heart-searching, we decided to keep the animals out of our garden if we could. This has meant a six-foot woven wire stock fence for the garden and the Rugosa rose patch and a wire and net cover over the blueberry quarter-acre in which we raise our cash crop.

Woven wire fences are good for only a few years—a dozen at best. After that they begin to wear or rust out. When our fence began to go to pieces, we were forced to replace it with new wire or find a substitute. Finally in Maine we went beyond wire protection and built 420 feet of stone wall around our entire garden area.

Certainly a wire fence is the cheapest form of protection against garden predators, but it does not give total protection. Beside our garden and west of the house, there is a small patch of lawn. Early on summer mornings, with the dew heavy on the lawn grass, a small army of slugs and snails would cross the lawn while the grass was wet, go through the fence into the vegetable garden, have their early morning snack and get back across the lawn to their regular quarters before we had finished our human breakfast and readied ourselves for the day's work.

At the worst of this invasion period we would take a container into the garden shortly after daylight, pick up to 500 slugs by count and dispose of them before coming in to breakfast. For a time this was a regular morning practice. As we replaced the wire fence by a five-foot stone wall we eliminated this threat to our supply of fresh green food almost 100 percent. The commuting slugs could not or did not climb over the wall.

Other items in the same category are perennial weeds—such as milkweed, thistles and witchgrass—which pass undeterred under or through a fence into a garden. A stone wall stops this invasion.

There are other arguments in favor of a stone wall. It is better looking. It is homemade. It lasts longer. On the other side of the argument, the stone wall takes much more time to put up and, if it is done at hourly wages, costs so very much more that the average gardener cannot afford it.

We have one strong argument in favor of our building stone walls: we enjoy working with rocks. In Vermont, stone was plentiful and well shaped for building. We found the same ample supply in Maine. A heap of fieldstone almost anywhere in New England will yield a surprisingly large number of what we call "builders."

In Maine as in Vermont we had a place off the mowing on the edge of the woods where stones were sorted and stored. Wherever we went we picked up "good" stones and carried them home, especially if they had one or more good "faces" or smooth sides. Lacking this virtue, they were classed as "uglies" and put in a pile for fill in foundations or roads. We seldom broke or cut stone, but picked from the stone pile the one that seemed most likely to fill the next place in the wall. As these stone piles grew, it became more and more evident that the time was coming when they simply had to be used.

During our first dozen years in Maine, nothing much was done with the accumulated stones. The piles grew in bulk and also in number. Finally, early in the 1960's we decided that the solution for the worn-out and rusting garden wire fence was a stone wall around the garden. We agreed that the project should be a part-time job. It could be our tennis and/or golf. Both of us preferred building with stone to playing tennis, golf or any other game. It was relaxing, in the open air, usually in sunshine. It was constructive and lasting. It was not the subject of an urgency or deadline. It would be our pleasant avocation.

The garden wall project had a unique advantage: it could be carried on for an hour or two, dropped for a day or two—or a week or a month—and then picked up again at our next opportunity.

The 420 feet of garden wall required as many feet of foundation trench. As the texture of the ground varied from one point to the next, this trench was dug to a depth that varied from 30 to 48 inches. With hardpan or rock ledge below, 30 inches was enough. With softer earth, 4 feet put us below the frost line.

The foundation trench was dug with mattock, pick, bar and shovel. The digging was done carefully so that we could use the sides of the trench to act as forms to contain the concrete and rocks while setting. On top of the foundation we put forms which would contain our stone wall. All of our forms were 18 inches wide and of various lengths. The forms were leveled and plumbed and braced strongly enough to hold the rocks and concrete.

Into these forms we put a bedding of concrete, then added the stones with their flat sides to the forms. Behind them we tamped the sticky concrete until it occupied every niche and cranny between the stones. Stone upon stone went into the wall, with at least an inch of concrete in between. When the concrete set, the form could be removed, or another form could be placed on top of the first one and leveled, plumbed and filled in its turn.

Building such walls depends largely on the available stone. Each of our walls begins as a pile of fieldstone, small and large, thin and thick. We set aside stones with 90-degree angles as corner stones, flat even stones as floor stones, stones with one flat face as wall stones, stones with no good faces as fillers or uglies.

During fourteen seasons of never-ending, varied, enjoyable open activity, the garden wall building went on. At times two or three people composed the work crew. At other times the number of participants was much larger. Most of the time we worked at it alone. The job was finished in the autumn of 1971, when Helen was 67 and Scott was 87. We give our ages to show that almost anybody at almost any age is capable of building a wall such as ours.

The total cash outlay for the garden wall was about $450, largely for the cement. If we had paid professional masons to do the job of building us such a stone wall, stone-faced on both sides and three to four feet underground, the cost would have been in the thousands of dollars. And we would not have had the pleasure and experience of doing it ourselves.

There were no fixed rules for the guidance of this fourteen-year volunteer enterprise. We just played it by ear, when and as we liked. Its duration outlasted any other activity of this kind in which we have ever participated. It was a collective enterprise too, in so far as many people who took part in the wall building learned some of the necessary techniques. Undoubtedly, however, the real value of the experience lay in three directions: first, its long-continuing occasional nature; second, the small numbers engaged on the job and the absence of any compulsion; and third, throughout the fourteen years on the job, although there was much discussion and exchange of opinion, there was little or no argument or bickering and, so far as we can remember, no quarrels. What an interesting and positive example of successful and constructive mutual aid!

We regret the need for fences, walls and other obstructions to free movement. If they must be built, we would hope that as many of them as possible could be solid, beautiful and provided, like this quarter-acre garden, with at least three gates, closed only when necessary, and available to anyone who has the wit to turn a wooden button.

"For the building of houses, townes, and fortresses, where shall a man finde the most conveniency, as stones of most sorts."

Captaine John Smith, Advertisements for the unexperienced Planters of New England or anywhere, *1631*

"The advantages of stone buildings are their great durability; their seldom wanting repairs; their greater security against fire; and their offering to the owners places of abode of greater comfort, both in cold and hot weather.... It may be thought by many that to erect such an one would be a great undertaking, yet it may be done without either great expense nor much difficulty. Hammered or chisseled stone is adapted to public buildings, or the houses of the wealthy, and is expensive; but comfortable, decent houses may be built with common stone, such as we would use for good field walls.... Their happy owners may live freed of that continual intercourse with the paint pot, the lumber yard, and the cut nails of all sizes and dimensions. A stone house substantially put up, will last three hundred years, and will require little or no repairs for the first fifty years."

J. M. Gourgas, in The New England Farmer, *January 1828*

"If I were commencing life again in the woods, I would not build anything of logs except a shanty or a pig-sty; for experience has plainly told me that log buildings are the dirtiest, most inconvenient, and the dearest, when everything is taken into consideration. As soon as the settler is ready to build, let him put up a good frame, roughcast, or stone house, if he can possibly raise the means, as stone, timber and lime cost nothing but the labour of collecting and carrying the materials. When I say that they 'cost nothing,' I mean that no cash is required for these articles, as they can be prepared by the exertion of the family."

Samuel Strickland, Twenty-seven Years in Canada West, *1853*

"I count it a duty to make such use of the homely materials at hand, as shall insure durability and comfort, while the simplicity of detail will allow the owner to avail himself of his own labor and ingenuity in the construction."

D. G. Mitchell, My Farm of Edgewood, *1863*

"These stone dwellings last forever, and need few or no repairs, so that money is well invested in them. Their quality does not deteriorate with time, like that of brick or wooden buildings."

Harriet Martineau, Our Farm of Two Acres, *1865*

Chapter 12

BUILDING STONE STRUCTURES

In Vermont, rock was everywhere, especially granite, excellent for building. We have found that Maine stones are as plentiful (though not as large and rarely granite) where we now live and are on the whole more colorful. Both Vermont life and Maine life were dominated for us by building with stone. Any day outside the deep-freeze season, after we had a look at the weather and consulted our date book, we would try to fit in some stonework, which we always found interesting, productive, creative and collective.

In Vermont we put up about a dozen stone buildings: a three-room cabin we built down the road off our land, to gain experience before we tackled our own home; a main house with connecting woodshed and sugar-packing room; a guest house and a workshop; a woodshed; a lumber shed; a garage; a greenhouse; a study for Scott on top of a huge boulder in the woods; and two guest cabins in the sugar bush—one occupied for years by Richard Gregg. With the log cabin we built to experiment on and to sell, we left quite a community of buildings when we moved to Maine after nineteen years in Vermont.

In Maine, from 1952 to date we have put up nine stone and concrete constructions. One of the first jobs was the spillway and concrete core for the earth dam that gave us our acre-and-a-half pond. Other undertakings early in our stay were several water tanks for garden and blueberry plantation, and a new tank for our bubbling spring. Another project was the 420-foot stone wall which encircled our quarter-acre vegetable garden. In the northwest corner of this garden wall we built a greenhouse, with the stone wall as a backdrop and sun reflector. To the west of the greenhouse, also along the garden wall, we built a stone garage.

Finally, we have been building three stone buildings on a small tract overlooking Penobscot Bay as a future homestead for ourselves. This big project was begun in 1972 by clearing the forest which had been in possession of the area long before Columbus discovered America. These acres had not been cultivated or built upon because they included a piece of low land lying between two stone outcrops which ran from the neighboring hills to the narrow, sandy strip of beach bordering the bay.

After much discussion, and some sounding and testing, nearly everyone decided the spot was too wet and swampy—not an auspicious building site, although having a superb view over the bay. An alternative site, which Scott favored, was to perch the projected stone buildings on a high section of ground also overlooking the bay. A building on this site would face north and could have no cellar unless one were blasted out of the rocky outcrops. It would also be difficult to get spring or well water up to the house. Helen favored the low area on account of the sunset view. She got her way.

The first finished project on the new place was a stone and concrete outhouse, completed in 1973. City-bred people may think it strange to launch a building plan by first constructing an outhouse. However, on any extended building undertaking far removed from ordinary utilities, putting up an outhouse the first thing is a matter of course.

It was built on a side hill with good drainage. The foundation was attached to a fairly smooth ledge with a pitch of 12 or 15 degrees. At the bottom of this pitch there is a wooden shutter hinged at the top and large enough to make the task of emptying the outhouse simple. The wastes slide down this incline and are easily shoveled out. If a generous amount of sawdust and soil is sprinkled by each user, the result will be a mature, odorless compost. The important thing is to provide enough earth and other absorbent material to handle the volume of waste deposited. We have found sawdust and wood dirt more effective than ashes.

The first time we emptied this particular outhouse (we had the same kind in Vermont), we were planting tulip and daffodil bulbs around the new house site. We placed the bulbs in depressions in the ground where we wanted a bed, and covered them lightly with compost. Then we went to the outhouse with shovel and wheelbarrow. We took out the material left from a summer's use (two wheelbarrow loads) and covered the bulbs with it. Then we put on more compost and topsoil. The bulbs benefited from the natural fertilizer, and the stuff was dry enough to leave the wheelbarrow and shovel clean at the end of the job.

Our outhouse was designed and built with hand-hewn timbers, a metal roof and a heavy Dutch door. Keith Heavrin, our near neighbor, and Scott did the foundation; Helen did all the stonework; Keith hewed the timbers and did the other necessary woodwork. It turned out to be a beautiful edifice and has been called "the prettiest outhouse in five counties."

We also have a composting toilet, or earth closet, in the main house—a Swedish Clivus Multrum. Outhouse and inhouse are useful in that they both return composted material to the land. The night soil will be used mainly under apple trees, as mulch, to stimulate their growth. As knowledge and practice of ecology grows, and water becomes a scarcer commodity, earth closets may take precedence of wasteful water closets.

Before we tackled our second stone building project on the new site, we had to put in roads. Ordinarily, that would involve buying many truckloads of expensive town gravel. We decided to look around for some on our own place.

The hill to the south of our new site rose high above the level on which we planned to do our building. It was a steep, ledgy outcrop, its rocky shoulders showing at various points among the young trees which covered the area with typical Maine coast vegetation, from the mosses and teaberry plants through alder and pineberry, white birch and soft maple, ash and black cherry to a few stately cedars and spruces.

Test digging on this rocky outcrop showed up a base of soft brown sandstone made up of small rock fragments the size of a hand to big rock masses. The rocky crag also had a covering of loam or forest soil so much sought after by gardeners for choice plants. Here was a fabulous treasure to be had for the taking: rich earthen pockets; rocks for the building; and stone chips just the right size for road foundation—millions of them.

Our roads were generally 10 to 12 feet wide. We established their levels, made provision for drainage and then laid down a solid bed of stones and chips from 8 to 15 inches deep. In wet places we began with large stones. On top of these stones we placed a layer of stones about the size of coconuts. On top of them went a layer of stone chips, finished off with a top-dressing of tidewater gravel.

This pattern of road construction is borrowed from the thousands of miles of hard roads that Roman engineers built in Europe, North Africa and the Near East about two thousand years ago. Some of the Roman-built roads and bridges are still in use. As with the Romans, no backhoes, tractors or heavy machinery were used in building our roads.

Results of this landscaping have yielded building rocks by the thousands, tons of coarse material for road building and topsoil for gardening. Best of all, the digging has uncovered a sheer rock precipice consisting of tier above tier of jutting rock ledges, set on the hill at various angles and providing a rugged backdrop for the stone buildings we have erected during the past three years.

The second unit of our stone and concrete building project was a garage-workshop-storeroom 25 feet wide and 50 feet long, with a metal roof painted brown to blend into the landscape. Why start with a garage and workshop? Why not do what is usual: get to the house first and let the lesser buildings trail along in due course?

In 1971 and 1972 there was little seasoned lumber to be had in the state of Maine. Trees stood in the woods on Monday; by Friday they had been cut and hauled to the sawmill. Within the following week they were being shipped out to construction jobs. We aimed to put our lumber in the open shed and leave it there to dry while we were building the house.

The garage-workshop involved digging 144 feet of foundation trench. The trench was 16 inches wide and at least 30 inches deep unless we hit bedrock, when we stopped digging. Most of the trenches were 30 to 40 inches deep.

When we reached a level at or below the frost line, we scattered 4 or 5 inches of fist-size rocks over the bottom of the trench and poured in a sloppy concrete mixture—one that would run into every nook and cranny. At the next stage we put in, on top of the small stones floating in concrete, a line of boulders as large as available and of a size possible to handle. Maximum width of the boulders was 12 inches, always aiming to leave a minimum of 2 inches between the side of the boulder and the wall of the trench, and with a minimum of 2 inches between boulders.

We continue this procedure until the trench is full, tamping and adding stones so long as they are covered by at least an inch of concrete. We aim to dig and fill 15 to 20 feet of trench at one time, then continue with another installment, until the foundation is completed up to ground level.

The foundation is left rough on top so that the wall of rock and concrete will have something to grip. Then comes the time to set the wall forms level and plumb and build the wall, which will be 12 inches thick. Our forms are 18 inches wide with a 2 x 3 spruce stud every 24 inches. They are held apart by 12-inch spacers and held together by a strand of telephone wire at the top and bottom of the form. The wire runs around the 2 x 3 studs if possible.

In building jobs as big as this, we like to begin with one corner, fill the form, then build a second corner. We then set forms and fill in the space between the corners, measuring carefully and frequently to keep the wall straight. We lay in metal reinforcing—especially at the corners. Thereafter there will be horizontal reinforcing each 9 inches as the wall rises—two lines of reinforcement to each 18-inch form.

At doorsill level, preparations are made to set in door frames, which along with the window frames can be set in either flush with the outside wall, flush with the inside, or midway. All three methods were experimented with on this building. Forms are shifted as we work up the wall until we reach plate level.

Keith Heavrin did the hand hewing of the timbers, the form work and general carpentry on this building. He stayed with us up to plate level. Then we had to put on the roof. Fred Dyer and his team did the job.

As the forms were removed from the walls, Helen did all of the pointing (which is filling in between the various rocks with a rich mixture of concrete). This is a one-man (or, in this case, a one-woman) job, as everyone handles the pointing mixture and trowel differently and the finished impression should be uniform in appearance. In all our various buildings and walls in

Vermont and Maine, this was Helen's special prerogative. She rarely let anyone (Scott included) handle the pointing trowel.

Who did the rest of the stonework? Just the two of you? No. Our work team consisted of the two of us if no one else was around; many particularly enjoyable days we quietly worked away together, with no outside help. But we get endless visitors and they usually want to help with whatever we are doing. The majority of them, boys and girls, men and women, have had little or no experience with such work, but they learn the ropes fairly quickly and many times become useful members of a smoothly functioning team. Or else they remember some errand they had to do in town and leave suddenly.

When we found how many workers were available on any set day, we paired them off. A couple of people made the mixes in the wheelbarrows and wheeled them to the job. One or two people were needed to keep the master mason (Helen) supplied with stone, which she insisted on handling herself.

With the roof on and the doors and windows in place in the new utility building, we had a tight, dry structure to store our piled lumber, our cement and our tools. On many jobs lumber is dumped and scattered about outside on the assumption that it will be soon used and that in the meantime no harm will be done if it is wet now and again. We proceed on an opposite assumption. On any construction job, we try to order our heavy supplies such as lumber and cement ahead. We like to keep them under cover and in order, stacked properly. Of course we take care of our tools, keeping them clean, oiled and out of the weather under some roof.

With the garage and workshop roofed in and a properly covered supply depot established, we were prepared in the spring of 1974 to make the preparations for building the house proper. It was to be two-storied, stone to the plates, with two bedrooms upstairs, each having a balcony. Helen designed the house to fit into the landscape. She wanted an alpine-type building with a broad sloping roof, creosoted heavy hand-hewn timbers, and pine paneling and bookcases in a large 20 x 30-foot living room with big windows looking over the bay. A Dutch architect friend drew up the first plans from Helen's initial drawings.

The first step was to dig a cellar under half of the projected house. Fred Dyer, a friend and local building contractor, looked the site over carefully and guessed that, although low-lying and wet, it could be used if extensive drainage were installed. With his cooperation we decided where the house cellar should be. Fred and his crew took on the job of putting in the cellar.

They excavated, drained, built the necessary forms and had transit-mix machines bring in the concrete and pour it. This was something new for us; every other cellar we had dug and built ourselves. However, this was bigger and wetter than anything we had tackled and we were glad to let Fred take charge. He let us help when it came to the actual pouring. The cellar when completed ran under the kitchen, hallway and bathroom. It is 52 feet long and 12 feet wide. We amateurs hand-dug and -filled foundation trenches for the balance of the house. We were ready to go on with the building of the house itself in the early summer of 1975.

At this point destiny moved into our building program with the appearance of a first-class cabinetmaker named Bretton Brubaker. A year earlier he had ridden a bicycle from his home in Ohio and turned up at our place, having read *Living the Good Life* and wanting to see what we were up to in Maine. He built a log cabin on a neighbor's camping lot and occupied it one winter. Then he got a job helping to refit an old schooner, the *Nathaniel Bowditch.* We visited the ship and saw some fine cabinetwork he had done. When it came to selecting a woodworker who could cooperate with us in building the new house, Brett was a logical choice, and he agreed to take on the job.

One of his first inspirations after seeing our plans for a simple, sturdy and longlasting house was a suggestion for the heavy timbers we wanted for the window and door frames. In Vermont we had plenty of big timber and had hand-hewed what we needed from our own stock. Here in Maine we had no big timber and would have to buy it.

Brett saw an ad in the local paper of a fertilizer factory being dismantled in a nearby town. The heavy mill timbers were being put up for sale. Brett went to have a look. He reported that they had many fine 6 x 12-inch oak timbers up to 24 feet long. The timbers had carried the mill roof studs and were bristling with 20- and 40-penny nails but were sound and serviceable. We bought two truckloads of the oak. Brett built an ingenious nail-puller and extracted the nails, smoothed up the timbers and used them to make all the door and window frames in the new house. With bar, winch and tackle he snugged the heavy timbers into place with little help. Like Helen, he wanted the fun of doing it all himself.

From then on, Brett was the genius of the place and did singlehandedly all the woodwork that had to be done. We and various visitors did all the stonework; Brett took care of the setting up and building of the forms. He proved an indefatigable and meticulous worker, a perfectionist of the first order. Perhaps we can best sum up his contribution by describing him as a one-man precision job, comprehensive and outstanding.

Two individualists really built the house. With Helen insisting on laying every stone, from outhouse through two large buildings to the high chimney, and with Brett tackling the careful carpentry and the cabinetwork alone, it might be termed a one-man, one-woman house.

Although others had a hand, Helen was in charge. She decided the overall plan, dimensions and layout. She designed and helped execute the inside furnishing and finishing. She handled every stone personally, choosing it, trying it out and putting it in its permanent place in the wall. Scott was allowed to mix the concrete in his favorite wheelbarrow, and dozens of unnamed helpers put in many days' work here and there on the house construction. But it really was Helen's and Brett's house.

A friend of Brett's, Forrest Tyson, was a retired electrical engineer and professor of electronics. Forrest had practiced and taught electrical engineering at the University of Hartford. He was good enough to help plan and set up the electrical installations of the new house. It was wonderful to work with friends on all these jobs.

We planned to heat this house and cook with wood. We put four flues in the massive chimney. A door on the north side on the house opened directly on the woodshed and had a storage capacity of eight cords. It was built open, with three piers, and covered by an extension of the house roof.

Earlier in this report we quoted a visitor as asking what we did for fun and enjoyment. There are many things we enjoy doing about the place: Helen prefers one and Scott another, but we both agree on one thing. Give us a good-size job of building with stone and concrete, with time and materials to do the work well. Now that the house is built, and we are living in and enjoying it, we look around us at piles of rocks still unused and wonder: "What next? Perhaps another garden wall? A green-house? A sauna?" Time will tell.

"The private buildings [in Virginia] are very rarely constructed of stone or brick, much the greatest portion being of scantling and boards, plastered with lime. It is impossible to devise things more ugly, uncomfortable, and happily more perishable.... The inhabitants of Europe, who dwell in houses of stone or brick, are surely as healthy as those of Virginia. These houses have the advantage, too, of being warmer in winter and cooler in summer than those of wood; of being cheaper in their first-construction, where lime, and stone, is convenient, and infinitely more durable."

Thomas Jefferson, Notes on Virginia, *1784*

"The evil in our architecture lies principally in this—that we build of wood. From this custom much immediate as well as remote inconvenience is to be expected. The comfort arising from celerity and dispatch does not make up for the numerous considerations of perishableness, want of safety, and call for repairs.... Bachelors only ought to build of wood—men who have but a life estate in this world, and who care little for those who come after them."

Anonymous, in The American Museum, *October 1790*

"I may remark here, in way of warning to those who undertake the renovation of slatternly country places with exuberant spirits, that it is a task which often seems easier than it proves."

D. G. Mitchell, My Farm of Edgewood, *1863*

"Many a farm of ample acreage is left to the rheumatic labor of advancing decrepitude.... There is no strength for repairs, no ambition for improvement, and no expectation of more than a bare subsistence."

Commissioner of Agriculture, Farming in New England, *1871*

Chapter 13

REMODELING OLD WOODEN BUILDINGS: DON'T!

DETAILS about the state of decay and dilapidation of farm buildings in the New England countryside are neither entertaining, constructive nor edifying. We write about them here because we want readers to realize what they have ahead of them if they tinker with old buildings. We lived in such houses for years, put up with their inconveniences and inadequacies, repaired, patched, reroofed, retimbered and otherwise spent valuable time and money on the projects.

We are aware that old buildings have an appeal for people fed up by the slapdash tawdriness and lack of grace and beauty which are so often met with in the "developments" that are going up all over the country. We agree in rejecting such graceless structures. But that is no reason for accepting and refurbishing old buildings that, no matter how patched, repaired and rebuilt, are still old buildings, even if the shingles and the paint are new.

We began homesteading in Vermont in two clusters of badly built wooden buildings, which were made of hit-or-miss materials and styled as the fancy of the moment or urgency or preoccupation had dictated. In a word, they had been thrown together. In each case there was a house built on a hole in the ground apologetically called a cellar. Each house was surrounded by a ramshackle woodshed, an outhouse, a chicken coop, a bathhouse.

When there had been a breakdown that needed repair, or signs of decay, the spots were plastered over or nailed or wired, patched or fastened back together with bits of lumber, rope, binder twine, chicken-coop netting or any other material that offered itself. Repairs had been haphazard, with no pretense at workmanship.

Of course there were exceptions, pleasant old houses, particularly in neighboring villages or towns, but we did not choose to settle in such. They were beyond our financial means at the time and outside the range of our interests. We wanted to live on a farm, on some isolated back road.

In Vermont, the Ellonen house where we started homesteading in 1932 was up the Pikes Falls road 7 miles from Jamaica, a village with a post office, country store, a bank and a church, and 3 miles from Bondville, which was little more than a country crossroads. In Maine, our Forest Farm is a little more than 2 miles from the post office and 6 miles from the nearest village (South Brooksville), 22 miles from the nearest town (Blue Hill) and 50 miles from the nearest city and airport (Bangor). On both of our homesteads in both states we started out with poor old second- or third-hand buildings of wood. In both places we tinkered with the decrepit places and finally built houses of stone for ourselves.

The Vermont house and barn needed roofing when we bought it. The house was ancient, with a poor excuse for a cellar, an outside toilet, and water from a pump in the kitchen. Our earliest building job in Vermont was to build a stone-walled, pine-paneled living room with a stone fireplace on the ground floor.

Some years later we acquired the place next door. It had a better cellar that included a spring of good water, but the house was in even worse repair than our original farm house. For the next two years we used the old house as a toolshed, carpentry shop, and shed for freshly sawed lumber direct from the mill. When we no longer needed it as a storage place, we tore it down and built a stone guesthouse and workshop on the site.

With the dried lumber and with rocks we had been gathering for years we built a new stone house against a split boulder on the edge of the woods about a hundred feet above the original wooden house. The split glacial boulder was 26 feet long and partly buried in the earth, from which it rose 9 feet above the surrounding terrain. We came across the boulder while cross-country skiing in the woods. The boulder was overhung with limbs from surrounding trees and smothered in brush and brambles. We paused in our skiing to examine the boulder more carefully, We decided that it was perfectly plumb and looked like the wall of a house. To that rock we attached the living room of our first stone Forest Farm.

The house near Hárborside, which we bought from Mary Stackhouse in the fall of 1951, was reported to be about a hundred years old. It certainly looked it and more. In its later years it had been used by woodsmen and hunters to house tools and supplies, and as a place to eat noon meals and occasionally stay overnight. When Mary bought the building it was in a bad state of repair. Tradition says that when she took it over, the back door was hanging by one hinge.

Mary was an artist of sorts with a modest amount of money to spend. Elwyn Dyer, who worked on the place as a carpenter, estimated that she had a thousand dollars to spend and house plans that called for ten times that amount. She had the work crew tear down most of the existing buildings, and pick out the best of the timbers, floorboards and siding from the wreckage. From this material and additional stuff they put up a five-room dwelling and a small barn-garage which used to be a boathouse. When Mary ran out of money she still had a spot for a fireplace that she planned but never built, a place for a woodshed that was never constructed and the beginnings of a garden that was never organized.

When we took over the place in 1951-52, the roofs of both house and barn were in poor shape; the floors downstairs were sagging; north winds blew right through the house, which was not insulated. For the next twenty years we spent time, energy and money repairing and remodeling the old wooden buildings. We did most of the work ourselves, so the money outlay was not large, but year in and year out we had to tinker, repair, replace.

We reroofed the house and garage, replacing wooden shingles with asphalt ones, and replaced some roof boards in the process. We replaced the wooden boardwalk porch back of the house with a stone and concrete terrace that connected house to barn. The barn-garage had been set on wooden corner posts dug into the ground. In the course of years these posts, in contact with the earth, rotted out. We jacked up the buildings unit by unit and replaced the wooden posts by stone and concrete piers with foundations below the frost line. The living-room floor was laid on 2 x 6 spruce joists rather too close to the ground. As a result the joists rotted and the floor settled. Elwyn tore up a large part of the living-room floor, replaced the joists and relaid the floor.

We tore a hole through the living-room south wall and built a stone fireplace. On the north and east sides of the house, at the second-story level, we put up a balcony for yodeling and to sleep on in good weather. After a dozen years this balcony became unsafe. It came down and we used the best of the rotted wood in the fireplace. At the end of all our work we had an old house.

Most home makers attempt to save money by making at least temporary use of an established building instead of devoting materials, time and money to the construction of a new one. It seems to us now that time, money, materials and effort can be used to better purpose elsewhere.

We have argued over the matter at length in our own family.

We have gone over the issues with friends and neighbors. Like so many problems, it can be discussed endlessly. Everyone has opinions and experiences. Not only have we gone over the pros and cons of old buildings versus new construction, but we have tried out one aspect of the problem after another for forty years.

Our conclusion is emphatic. Do not spend time and materials in reconditioning old structures. Nine times out of ten it is better, cheaper and in the long run more satisfactory to build new and with the best materials available.

Building a shelter for self and family is not a fly-by-night affair. You will probably live many years in the new quarters. See to it that the design, the materials and the workmanship are the best obtainable. Economy, standards of workmanship, aesthetics and historical experience all press for excellence.

Our advice on the remodeling of wooden buildings would be something like this:

1. The cost of repairing and remodeling old buildings is about the same as the cost of new buildings. Probably in the end the cost will be a bit more. Certainly in our experience it will not be less.

2. In remodeling, if some materials are reused in the new structure, particularly if they are of wood, they will rot out much sooner than new ones.

3. Unless the overall pattern of the old building was letter-perfect and repeated letter-perfect in the new one, the result will be a clash of styles. The structure will lack uniformity and integrity—will be neither old nor new.

4. Anyone who builds has ideas and concepts which are part of the personality and should not be lightly cast aside. Unless the remodeler and rebuilder is an antiquarian, doting on the old, believing in the old and following it slavishly, we would recommend: Build according to your own personality; do not adapt to someone else's life style and tastes.

If you are not tied to the past, if you are mechanically inclined and creative, build new every chance you get, with the best materials that come to your hands, and do it yourself. Work out and develop your own designs. As you proceed and are painstaking, each new venture will be a chance to express your own ideas and to improve your skill as you strive to work out the very best product of which you are capable. Following such practices you will create and learn—learning and creating at the same time.

photo: Richard Garrett

"He is wyse, in my conceyte, that wyll have, or he do sette up his howseholde, two or thre yeares rent in his cofer."

Andrew Boorde, A Dyetary of Helth, *1542*

"Every work for the next day is to be arranged, whether for fine or rainy weather, and the farm-books to be made up for the transactions of the past day. Besides these, he should have another book, for miscellaneous observations, queries, speculations, and calculations, for turning and comparing different ways of effecting the same object.... Loose pieces of paper are generally lost after a time, so that when a man wants to turn to them to examine a subject formerly estimated or discussed, he loses more time in searching for a memorandum, than would be sufficient for making half a dozen new ones; but if such matters are entered in a book, he easily finds what he wants, and his knowledge will be in a much clearer progression, by recurring to former ideas and experience."

Arthur Young, The Farmer's Calendar, *1805*

"I would encourage every family to live within their means. If there be a way—and such a way there certainly is—of living as comfortably and happily, on very small means, as we now do on much larger ones, it is certainly desirable to know it, especially in times like the present. 'But suppose the means are very small, what then?' Why, then, live within very small means."

William A. Alcott, Ways of Living on Small Means, *1837*

"No one will deny the importance of urging rich and poor alike, in the present state of things to try and economise the fuel and food which they may have at their disposal. The sooner we make up our minds that what we regretfully speak of as the 'good old times' with their good old prices will never come again, the sooner we shall cease to look fondly back on a cheaper past, and brace ourselves helpfully and bravely to face the increased cost of the necessaries of life."

Lady Barker, First Lessons in the Principles of Cooking, *1886*

"Everything depends both on what has been and what is to be. Which suggests the need for a good system of records.... It is not the fellow weeding the onions in July who is getting the most out of his farm; it is the man who, in January, is planning what kinds and quantities of onions to plant next Spring.... It may be possible to run a farm well without the help of a carefully worked out plan, but if it is I have never seen it done."

Henry Tetlow, We Farm for a Hobby and Make It Pay, *1938*

Chapter 14

PLANS, RECORDS, AND BUDGETING

HUMAN beings are persistent planners and record keepers. Stone Age men chiseled their records on massive boulders. Those who came later used horn, wood, baked clay, animal skins and vegetable fibers. Whatever the medium on which men have listed their prospects, outlined their purposes or detailed their plans, humans have recorded the past, surveyed the present and made their proposals for the future as a matter of record.

Written records have a multitude of uses. They are as important to the gardener and the garden as they are in most other fields of human interest and endeavor.

Successful gardening begins with a survey of the proposed garden spot—an evaluation of its possibilities and limitations. It continues with a freehand outline of the project. Soon after it is put on paper, the freehand sketch is finalized by putting into your garden book a working drawing, still in free hand, but outlining the general garden project.

If possible, the garden should slope gently toward the south. Certainly it will aim at maximum exposure to the sun, and a minimum of shadow from trees or buildings or even from taller garden crops. If tall corn or pole beans or other crops are contemplated, they should be arranged in a manner that will throw the least possible amount of the garden area into shade.

A second important consideration is crop rotation. With rare exceptions the same garden crop should occupy its place in the garden for only one season. In the second and third seasons, different crops should occupy that space.

Irish potatoes do better if planted on a fresh green sod each year. We arrange this by having four plots side by side, each plot 15 x 30 feet. We begin the rotation by planting potatoes in plot 1 which, ideally, was occupied by a green sod the previous year. Meanwhile plots 2, 3, and 4 are planted to other crops.

The second year, potatoes go into plot 2. Plot 1 meanwhile is planted to squash. The third year, potatoes go into plot 3 and squash into plot 2; plot 1 is planted to grass and clover. The fourth year, potatoes are in plot 4. Squash goes into plot 3. Plot 2 is bearing a ragged green sod and plot 1 has a rich green sod. The fifth year, potatoes take over the green sod of plot 1. Squash occupies plot 4. Grass seed is planted on plot 3, and plot 2 is allowed to develop a green sod of its own.

Here is a minor crop-rotation sequence. Each fifth year the same crop returns to its 15 x 30-foot land strip. Between any two returns there are three years devoted to particular crops: one to squash, and two to green sod.

In our garden, as presently laid out, there are two paths which divide the garden roughly into four segments. Each year one of the four segments produces tall peas, a second produces pole beans, a third produces cucumbers, melons and tomatoes, and a fourth, small truck. Each year these four crops move on to a new garden location, rotating the crops into a new position.

Gardens have a fifth segment—semi-permanent crops, such as asparagus, artichokes, rhubarb, strawberries, which continue without any annual rotation for a number of years. Strawberries stay in one place for two or three years. Rhubarb and asparagus beds, once established, last a decade or more. Semi-permanent crops should be grouped together and so located that they interfere as little as possible with garden routine.

Biodynamic gardeners advocate bed culture instead of row culture. Most organic gardeners in the United States lay out their gardens in sections and plant almost everything in rows. Exceptions are made in the case of pole beans, corn, cucumbers, potatoes and squashes, which are planted in hills. Using hand- or power-driven machines, gardeners are able to go over their rows in short order. Weeds are eliminated and gardening correspondingly simplified.

Bed gardening is quite different. Beds are raised several inches above the surrounding garden area. Sometimes the beds are boarded or planked along the sides. Generally they are made by heaping up the soil to form the bed, making the bed itself so narrow that it can be worked from paths on both sides without treading down the soil of the bed.

This is not a modern invention. Chinese and French advocates of intensive gardening have planted in beds for centuries. The French method is usually quite small in scale; the Chinese often extend their beds for hundreds of feet across relatively level land.

Generally they are about a yard wide, skipping a space to provide a walkway between, then heaping up a second raised bed parallel to the first. The garden thus becomes a series of raised beds, uniform in width, with narrow walkways between them, producing drainage ditches for wet land and irrigation ditches for dry land.

The raised beds are planted and serviced by pairs of workers, one working along the ditch on one side of the raised bed, the other working the opposite side. By bending and reaching, two workers cover the entire bed, keeping their feet in the path, without stepping on the bed. Any soil, but especially a clay or gumbo soil, thus treated, remains light and flaky, making the pulling of weeds easy and allowing the plant roots to penetrate the soil.

The benefits are obvious. In our previous row-culture gardens we have packed down the loose soil by rolling it, beating it down with a spade or even walking on the seed row after it was planted. With bed culture the gardener keeps his feet in the walkways and does not pack down the seed rows at all, but makes every effort to keep the soil open to sun and air and capable of absorbing a maximum of moisture.

If the beds have been properly prepared and the soil is flaky and friable, sunlight and air penetrate the soil freely, water is absorbed into the soil as it falls with a minimum of evaporation or runoff and thinning and weeding are greatly facilitated.

Beds may be planted crossways in short rows, or longways in long rows, or the seed may be broadcast, raked or rolled in, and give the workers a chance to thin the plants to proper distances and at the same time to pick out the weeds. The most impressive result of bed gardening is the relatively high yield per square yard of garden due to the greater density of plants.

In bed gardening, as in row gardening, it is essential to know what has been planted and at what date. We meet this need by having a number of plant markers—wood, plastic or heavy waterproof paper on which we note with a ballpoint pen the date, variety and any other necessary information.

The same information, noted down in a garden book, provides the gardener with a record so that a glance at the book will tell the story of that particular bed, row or hill.

Like any other complex and confusing situation, a garden book (or a box of index cards kept up to date), enables the gardener to proceed from day to day and season to season with enough information to know what is happening in various parts of the garden and for like periods in various gardening seasons.

One can go further, making a detailed garden plan in advance and checking off the various items as the plan is carried out.

This brings us to an item of the greatest concern to homesteaders and any other experimenters—the role of bookkeeping (a) in making plans, (b) in carrying the plans into execution, (c) to establish the limits within which plan and fulfillment must operate and (d) to balance accounts at the end of the operation, to show at a glance whether the operation has involved a profit, suffered a loss or broken even.

Behind every successful enterprise there is not only a plan but also a series of preparations that will make the success of the project more likely. A decision to set up a homestead is not a casual affair but should be the outcome of careful thinking, a series of firm decisions and the determination to see the project through to a satisfactory conclusion. The first three years are the most difficult and critical. It takes at least that length of time to try out and check on the requirements and the possibilities.

Anyone who attempts to abandon a market economy and move into a use economy faces a period of transition that will certainly eat up months or perhaps years, during which the security of the old economy is absent and the hopes and promises of the new economy have not yet materialized. This gap must be filled by an amount of working capital sufficient to provide goods and services during the transition interval. Anyone planning to make such a shift must be prepared to provide the necessary goods and services for existence during the interval.

Most of our friends and acquaintances who have shifted from a market economy toward or into a use economy had laid aside enough to provide food, clothing and shelter for at least one transition year. Some prepared for two years or even three.

Others have arranged to keep working part-time for months or years of the transition. In the case of groups larger than family size, some of the members have continued for years on regular jobs in the market economy, contributing the surplus of their earnings to the group treasury.

The first requirement is fellow workers. Only an exceptional individual—a confirmed hermit—can set up a successful homestead alone. A would-be homesteader should have a tried and experienced partner or partners. Perhaps we should say, the more cooperators the better, but at least one who will stick to the project and see it through.

A homestead, to be successful, must operate day and night for 365 days in the year and for several years before it is "out of the woods." It is a big assignment for any individual to be continuously on duty and responsible week after week, year after year. Both the responsibility and the necessary labor should be shared. Only the "right people" will succeed in homesteading.

Experience is a good teacher. If one or more of the prospective homesteaders are tried and true, so much the better because they will have learned some of the skills that are necessary for such a project.

A homestead presupposes a piece of land as the base for operations, enough land to make success likely or even possible. The land should be suited to the needs of the proposed homestead. If possible it should be owned and unencumbered by mortgages or other obligations.

The homestead will require working capital—that is, a supply of cash or at least credit sufficient to carry the enterprise for at least two, and better still three, years without any net income from the project itself. Of course there will be some income which can be added to the working capital, but that is uncertain and not to be counted on. The initial working capital should be sufficient to carry the project through an extended trial period.

During that trial period the project should acquire the needed tools and equipment, paid for out of the initial working capital.

Skillful bookkeeping is another requirement. Someone on the project must be responsible for all of the necessary paperwork, including the maintainance of a cash balance sufficient to cover all necessary and likely expenses.

If two or more people are engaged in the experiment, they should be seriously and honestly determined to make the experiment succeed, to think through each day, week, month and year. In a word, each step should be anticipated, put on paper, discussed, modified if necessary, agreed to and then carried out.

Each day, week and month should have a working program, put down on paper, and referred to constantly as a guide and determining factor. The whole project might be referred to as a "three-year plan."

Most homesteaders have sharply limited resources. In that case, if they hope to succeed in their homesteading adventure they would be wise to follow the old adage: "Pay as you go." If the homesteader is wise, he will do his utmost to keep his spending inside his income and to avoid borrowing and interest slavery as he would any other evil.

The homesteader should distinguish between current expense and capital outlay. The budget for each year should contain a balance sheet, with current receipts on one side and current expenses on the other. Every effort should be made to end each year with a surplus or profit rather than a loss or deficit.

The budget surplus, like any other profit, can be consumed for personal needs and wants, or it may be put back into the business by buying new tools or other equipment or by laying in stacks of raw materials.

Sugaring in Vermont yielded a surplus or profit each year, but it required capital spending to replace worn-out equipment. Each year sap buckets were damaged, piping was replaced and new tools were purchased. Each year we set aside a part of our sugar-business surplus to cover such depreciation.

We also decided to insure the sugarhouse and its contents. Sugarhouses were littered places, fires in the evaporator were hot and many syrup and sugar makers suffered serious fire losses. We went to an insurance agent. Oh yes, he would insure our sugarhouse and contents for $500, but the risk was great and the premium was high. The policy would cost us annually 25 percent, or $125.

We thanked the insurers and went our way. Each year we took $125 from the sugar-business surplus and put it aside. At the end of four years we had $500. With this money we built a second sugarhouse, installed a duplicate set of sugar tools and equipment so that in case the first sugarhouse burned during the sap season we could finish out the season in the second sugarhouse. Incidentally, if we had a big run of sap we operated both sugarhouses at the same time and were able to double our output in a single good sap day.

United States big business has installed methods of bookkeeping—cost accounting, depreciation, special funds and installations—that gobble up surpluses and avoid tax collectors. One thing we small fry can learn from big business is to keep careful records, anticipate the future and avoid trouble in one of its many forms.

Plans and records can be as important to little business as they are to big business. Well-kept records tell the story of any enterprise and in the long run become the basis from which the history of the enterprise can be written.

photo: Richard Garrett

"Sit down and feed, and welcome to our table."

William Shakespeare, As You Like It, *1599*

"Every man shall eat in safety, under his own vine, what he plants; and sing the merry songs of peace to all his neighbours."

William Shakespeare, Henry VIII, *1612*

"I had more visitors while I lived in the woods than at any other period of my life. I have had twenty-five or thirty souls, with their bodies, at once under my roof....But fewer came to see me on trivial business. In this respect, my company was winnowed by my mere distance from town."

Henry David Thoreau, Walden, *1854*

"Over the river and through the woods to grandfather's house we'll go."

Lydia M. Child, Thanksgiving Day, *1870*

"Digging was not what any hand yearned for. Some are born to dig; others have digging thrust upon them."

E. F. Green, A Few Acres and a Cottage, *1911*

"Doorbells are like a magic game, or the grab-bag at a Fair—You never know when you hear one ring who may be waiting there."

Rachel Lyman Field, in Taxis and Toadstools, *1926*

"She left no little things behind, excepting loving thoughts and kind."

Rose Henniker Heaton, The Perfect Guest, *1930*

CHAPTER 15

VISITORS AND HELPERS

HUNDREDS of people came to see us and our farm in Vermont. The thousands of young people who now come to our farm in Maine are the same type of seekers. They have heard or read about our Forest Farm and are curious to learn what it has to show or teach. They are ready for anything that makes an idealistic appeal and that is fairly far from standard community practice. They are unattached except in the very limited sense of selective mating. They are apolitical, impatient of restraints—especially when governmentally imposed.

Increasingly they are turning their backs on a world community that has tolerated war and is preparing for the contingency of one in the future. They are ardently in favor of peace in a broad sense, but are not ready to accept a commitment to any organization that works collectively for the cause. Almost universally they favor "freedom": that is, the pursuit of their personal goals and fancies. They are not joiners and generally not members of any group more specific than is implied by the adoption of a specific diet or the practice of some yoga exercises.

They are wanderers and seekers, feeling their way toward an escape from orthodoxy and superficiality, with the nervous dissatisfaction that characterizes people who do not have a home base in any real sense. Perhaps they can best be described as unsettled. Never before in our lives have we met so many unattached, uncommitted, insecure, uncertain human beings.

There are, of course, those serious few who are consciously and conscientiously working toward an ideal in which they believe and to which they attach themselves. They are definitely looking for a niche in which they might play a more effective part in helping to develop a new and better life style.

Many young couples, singles and groups came to us wanting land. They had limited means. Most of them had traveled from coast to coast looking for a place that they could afford and that would offer possibilities for a satisfying alternative way of living. They were homeless between two worlds.

We had a hundred acres or more (our deed read 140 acres, but it turned out to be a good deal less, and at the moment we have only 26 acres left). We decided to share what we had with the right young people. We sold a large slice of land to Susan and Eliot Coleman, a promising young couple, for what they could afford to pay; another slice to Jean and Keith Heavrin; and later, the tip-end away from the water to Greg Summers. All were prepared to homestead. All cleared their land and put up the necessary buildings.

The Colemans went in for raising vegetables and fruit as a cash crop and made a great success in market gardening. The Heavrins built a fine house and deep cellar, added a large pond, went in for animal husbandry and did construction work as a source of cash income. Greg Summers, who had an art background in Wisconsin, raised a good garden in the woods and took printshop jobs to bring in the necessary cash.

A real problem was the large numbers of singles and couples who wanted to spend a few days, a week or a summer working with us, learning and practicing gardening and building. We received hundreds of letters from them asking if they could come and work and camp. Where could they stay? On part of the Colemans' cleared land we put some cabins and heating and cooking arrangements and provided the bare necessities for backpackers and helpers. Each open season the campground was utilized.

The young people who gathered there were strangers to each other but readily found things in common. The shifting population worked wherever and whenever they felt disposed. Meals and garden produce were provided where they worked. They had musicals around the campfire. They attended our Monday night meetings where everything from compost to communism was discussed weekly. They went to square dances in the neighborhood. On the whole, it seemed as though they had worthwhile and happy times. It was a free and easy, loosely knit community formed of people who came and went. Some of the campers and workers became our very good friends, corresponding and coming back at every opportunity.

Parallel activities are going on in various parts of Maine and elsewhere in New England and the United States, where by various means land is being secured and homesteads, collectives and communes are being established.

We do not remember having one black youngster come, or one daughter or son of a coal miner, nor do we recall young people whose parents worked in textile factories or steel mills. We had daughters and sons galore of merchants, of doctors, teachers, lawyers, bankers and public officials—people who had been born into affluence of a sort, raised in comfort if not pampered in luxury. They had had little or no real work experience. They might have washed dishes, run errands, sold merchandise or taught school for short times, but they had stuck at nothing for long. Occasionally we had a visitor who had worked on a farm and knew his or her way around. Some few had real skills. They were among the exceptions.

We realize that the few thousands of young people who come

our way are not necessarily representative of American youth in the decades following war's end in 1945. It is quite possible that the young people who stop by at Forest Farm are leftovers who do not fit into the current picture and are deliberately seeking another life style. They are part of the process of self-selection during which those inclined or ready to line up with the Establishment have made their choices and are securely part of the typical American scene. In one sense our visitors are a motley crew. Through trial and error the sheep are being separated from the goats—or, the goats from the sheep.

Before we moved from Vermont to Maine, the trickle of visitors had become a stream. During the next years in Maine it became a flood. By the 1970's the number of visitors by head count has ranged between 2000 and 2500 in the course of a year. It often reached dozens in a day.

How do we cope with this visitation? By a single and generally followed rule. When visitors come unannounced we continue doing what we had planned to do for that day and that time of day. If we are writing, we go on writing. If we are working, we go on working. This is not to the liking of those who want to sit around and talk about this or that. However, we have to keep on our jobs in order to get anything done.

Many visitors are willing, eager and enthusiastic. Some few prove to be experienced people who know how to handle tools and what to do with them. Some make real contributions to the projects on which we are engaged.

We had a few glacial boulders that had found their resting place on our land. The largest of these boulders must have weighed between a quarter and half a ton. It was in our way and we decided to move it about thirty feet. One of us alone could not budge it, even with a crowbar. Five of us could handle it only if we worked as a team, following the directions of a temporary leader or foreman. With an explosive or a series of drilled holes and wooden lugs, we might have broken the rock, but only by spending a long time on the job. In any case, we wanted it in another place. Finally, with bars and chains and the help of a four-wheel truck, we shifted it to its new dwelling place. Teamwork had paid off.

Necessarily, with the numbers of visitors ranging from one or two to a dozen or a score in a single day, our helpers had to be grouped together in one or more work teams. Of course, there were always drop-ins and dropouts. But if we had reason to believe that two, three or four experienced and reliable people were to be on hand, we organized as many crews as we could handle effectively on that day. We divided up workers according to inclination and experience. Most crews consisted of three to five people under the direction of one or two experienced workers. With these crews we could dig foundation trenches, pour concrete, cut brush and above all build roads. People who had skills exercised them. Others of less experience learned.

If the two of us were alone, there was no problem. We could work in the house, in the woods, in the garden, or at building. If there were more people we tried to adapt to their number. Building was not their forte; masonry we did not dare let them handle. On gardening we could usually use only a few experienced people. Work in the woods with sharp tools was dangerous if not handled by knowledgeable folk. So we usually came down to clearing land or building roads when there was a crowd. They did not always like this; they would rather sit and talk, but we went at the job anyway, with those who were willing.

Perhaps the hardest work they did was on the craggy rock pile. They called themselves "the chain gang" but they sang and worked with a will, with occasional rests and lapses. Some tenderfoot visitors, facing a rocky hillside cleanup job for the first time in their lives, were put off by the roughness of the work. Some were unprepared to get their hands into common dirt, while others were attracted by the archaeological excitement of uncovering down to bare rock. To some it was tedious, to others exciting and fascinating.

We laid out a road network and proceeded to build the roads: with large stone, with small stone and gravel on top. Where did we get this material? From the beach on our cove and from the rocky outcrop beside our house site. We attacked the cliff with axes, picks, bars, mattocks, forks, shovels. We cut off trees, shrubs, brush. We dug out roots. We scraped the woody surface with trowels, carefully collecting, from the pockets of forest, soil which we saved for landscaping or garden. The feminine members of the group usually picked this easier sitting-down job, although there were sturdy girls who preferred to work fiercely with mattock or axe, or wheel the heavy loads of rock or miscellaneous refuse away from the working face of the cliff to the roadways being built.

Large rocks went to their allotted piles to use for terraces if they had good faces, to be used in foundations if they were uglies. Medium stones were carefully sorted into builders, corner stones and uglies. The rough balance was shoveled into the wheelbarrows and wheeled to the road we were constructing around the buildings.

We are writing about our visitors during the period of construction work on the new house and grounds. During the last five years we have worked off and on at this time-consuming project, fascinating in its way, and we are still at it. It has involved the participation of neighbors, friends and chance itinerant help, some of which has become semi-permanent and paid. We and they, we hope, have enjoyed our contacts together. There has been an apprentice-learning side to our endeavors, a sociological-communal side, a constructive side and an entertainment side. The results generally were educational, healthy, beautiful and useful.

We have not yet mentioned the culinary side of our day's work, which was Helen's domestic contribution. After a morning's work in woods and garden or at road or building work, we were called at noon. We gathered our tools, cleaned them and put them in place. Then we trooped in to lunch. Weather permitting, we ate outside on the stone patio that adjoins the kitchen. Chance visitors who were on hand at noon were also invited. Each luncheon guest received a wooden bowl and wooden spoon. Soup was dipped out, with always seconds or even thirds available in the tureen. Season permitting, green leafy vegetables were on the table, and sprouted grains. Bowls of uncooked rolled oats, oil and raisins ("horse chow"), boiled wheat, millet or buckwheat were served. These cereals were eaten with an amalgam of peanut butter and honey called "Scott's emulsion." "Carrot croakers" were popular journey cakes with a grated carrot base. Apples usually topped off the plentiful though simple meal.

The number of hungry mouths at these lunches ranged from half a dozen to over twenty on some days. We tried to serve an abundance of simple, fresh, nourishing vegetarian food: food that was ethically and dietetically sound and sufficiently filling to satisfy people, mostly young, who had spent four hours in hard physical work. With ravenous appetites, the spread usually disappeared in short order. Officially the meal lasted for an hour. If there was interesting talk and no other pressing work, people stayed on longer.

This brief review of our experience with two generations of helpers and associates during the last few years hardly does justice to what is really a period of transition and transformation. Changes are taking place, deeply affecting the young, and we—with our lives almost over—observe them with mixed feelings. We believe the groundbreaking work we have done with our own homesteading and the missionary end of it in letting people come and observe has been worth while through the years. We would like to continue to have interested visitors drop by. We are glad to have willing helpers when they want to assist in any of our projects. But by 1976 the handling of hundreds of visitors had become so serious a problem that we put up a sign at the entrance of Forest Farm stating that our mornings were our own and that we could accept visitors from 3 to 5 in the afternoon only. In 1978 we declared a sabbatical (our first) in order to get to some necessary writing (this book included) and we would see visitors only with advance notice. We regretted this seemingly inhospitable step but had to take it in order to get any amount of consecutive work and writing done.

We would like to take every opportunity to help young people in their serious search for a life style that would make sense to them. All are possible recruits for a general effort now under way to stabilize and improve man's earthly living space. For a long period we have done our utmost to raise popular interest and determination to the action point. We hope to continue to do so as long as our energies last. As we near the century mark of our lives we find we must limit our contribution to part-time.

"Were it in my Power, I would recall the World, if not altogether to their Pristine Diet, yet to a much more wholsome and temperate than is now in Fashion."

John Evelyn, Acetaria, *1699*

"A very spare and simple diet has commonly been recommended as most conducive to Health."

Dr. W. Kitchiner, The Cook's Oracle, *1822*

"All men eat fruit that can get it; so that the choice is only, whether one will eat good or ill; and for all things produced in a garden, whether of salads or fruits, a poor man will eat better that has one of his own, than a rich man that has none."

John Loudon, An Encyclopedia of Gardening, *1822*

"With respect to luxuries and comforts, the wisest have ever lived a more simple and meagre life than the poor. The ancient philosophers, Chinese, Hindoo, Persian, and Greek, were a class than which none has been poorer in outward riches, none so rich in inward. . . . Simplify, simplify. Instead of three meals a day, if it be necessary eat but one; instead of a hundred dishes, five; and reduce other things in proportion."

Henry David Thoreau, Walden, *1854*

"Wholesome food and drink are cheaper than doctors and hospitals."

Dr. Carl C. Wahl, Essential Health Knowledge, *1966*

CHAPTER 16

WHAT WE EAT AND WHY

WHAT IS man's natural food? Surely not the mishmashes and inadequate, denatured items displayed on the shelves of supermarkets. They are bad-natured, instead of good-natured, foods. The artificiality of their worked-over substances should be apparent to the most casual and conventional food shopper.

There is hardly anything natural in the food stores anymore. The grains are puffed, pounded and supersaturated with sugar for boxed breakfast cereal. The meats are injected with hormones, colored and full of antibiotics. The juices are filtered, flavored and carbonated. The overprocessed canned fruits are sweetened beyond measure. Sugar is also included in the packaged soups, soft drinks, salad dressings, sauces, dessert mixes, ketchup and even peanut butter. The raw vegetables in the stalls are sprayed and dusted; the raw fruits gassed and coal-tar dyed; the ice creams notorious for their heavily sweetened, artificially flavored compounds.

Restaurants and food bars where so many go for their meals offer quick-service items that may have been standing on the stoves for hours, losing vitamins, and before that sitting in warehouses and trains and closets for weeks and months. The foods that are bought and eaten in the United States are handled and handled and handled again. By the time the hungry diner is served, his dinner is doctored and tampered with and far from natural.

Where do we turn for good, healthy, nourishing, unadulterated food? To our own gardens, of course, and our own kitchens. Grow your own; cook your own—or, better still, eat your own food raw. The best and simplest food in the world is unprocessed, picked by yourself from the garden or the trees or the meadows or the woods and eaten raw before the vitality has fled. This is vital, this is nourishing, this is good-natured food.

Five minutes away from contact with the earth or mother plant is enough to start the wilting disintegration process in lettuce and other delicately leaved greens, and deterioration in the more solidly constituted vegetables and fruits.

As for food preparation, the less and the simpler the better: the better for the housewife in time spent, the better for the food itself in vitamins saved, and the better for the body in vitality added. Processed foods are those that have been cut, peeled, cooked, frozen, pickled, pasteurized, spiced, colored, smoked, flavored, chemicalized and otherwise adulterated.

We never have gone in for denatured food and have kept away from store food whenever possible. Eating out, especially in restaurants, is never a pleasure for us. One winter we were in Asia and the Near East for almost five months. During that time we never entered a restaurant to eat. We frequented the local markets, bought greens and fruits and cheeses and ate in our rooms. The food we consumed was as fresh, as simple and as unprocessed as we could find. It was far healthier and safer and cheaper than eating at restaurants.

We can be and are largely self-sufficient in food. Self-sufficient means that we can feed ourselves. During half a century of gardening there has never been a time when we have lacked a supply of organically grown produce. This food comes directly from the garden for a large part of the year. During the late fall, winter and early spring we have three additional sources of supply. (1) It may come from the sun-heated greenhouse where we grow lettuces, parsley, radishes, leeks, kale, spinach. (2) It may come from our root cellar where apples, potatoes, carrots, onions, beets, rutabagas and other root vegetables are stored in bins of autumn leaves. (3) It may come from our stock of bottled soups and juices and applesauce which we put away during periods of surplus production and use when the garden is in deep freeze.

We use several grains in our diet, which we do not grow on our farm: wheat, oats, millet, buckwheat, rice, corn, are bought from the local coop. With the exception of apples and pears, we have no local source of tree fruits. Berries grow well with us: blueberries, strawberries, raspberries. Grapes we have found need a longer season. We buy and use vegetable oils: olive, peanut, safflower. We buy citrus and bananas in a local wholesale market. With these noted exceptions, we come as near to our ideal of food self-sufficiency as seems presently practicable.

From ordinary grocery stores or supermarkets we may buy occasionally citrus, bananas, avocados, cheese, yogurt and sour cream (though we are turning to tofu and soybean milk instead of dairy products).

Bread is a time-consuming product and so easy to glut on, particularly when laden as it usually is with butter, jam, cheese and other things. Instead of spending hours making and baking breads from manipulated dough, we eat our grain unground and unkneaded. Bread is so easily swallowed unchewed; grains are grittier and require mastication, which is good for the teeth, gums and digestion. If tough and hard, like raw wheat berries, we soak a couple of handfuls overnight, and simmer on the stove for a morning. If a softer grain, like millet or buckwheat, we cook for less than half an hour. Rolled oats we eat uncooked.

Put on the table, hot or cold, these grains are eaten with a bit of oil and sea salt, or with an amalgam of honey and peanut butter, and sometimes chopped-up apples. We do not use white flour in our home and rarely any flour, grinding our own grains when we want some.

We buy in bulk through our local cooperative or health-food store and always have stocks of staples on hand, stored in barrels or tins. We could almost qualify for a health-food store with our supplies of wheat berries, rolled oats, buckwheat groats, cracked wheat, oat flakes, soybean flakes, barley, lentils, millet, rice, cornmeal, popcorn, dried beans and peas, mung beans, raisins, prunes, dates, peanuts, peanut butter, almonds, sunseeds, herbs, honey and oil.

We rarely buy canned or frozen vegetables and fruit, or snack foods, and never the cake mixes, puddings, desserts, pizzas, cookies, pastries and instant goods that crowd the shelves of food shops.

We try to avoid sugar and refined sugar products, substituting instead honey, maple syrup, molasses and sweet fruits such as dates, raisins, prunes.

Living in a climate not adapted to a year-round fruitarian or raw diet, we adapt and simplify as much as possible. The basics of our diet are simple: our own herb teas and fruit for breakfast; soup and grains for lunch; salad, one cooked vegetable and some applesauce for supper. Day in, day out, these are our meals. There may be a few additions occasionally—such as sunflower seeds or nuts to the breakfast; or tofu or cottage cheese to the supper—but these are the fundamentals.

There is an easy ritual to our mealtimes, which are eaten at 7, noon and 6 o'clock. On a bare wood table we set wooden bowls, wooden spoons and chopsticks. Soup, salad or dessert is all served in the same wooden bowl. This simplifies dishwashing. We are not quite down to Diogenes' standard of simple living:

when he saw a boy drinking out of his hands, he threw away his only utensil, a cup. Nor do we adhere to Gandhi's determined silence at meals.

Our meals are sociable times when conversation can range from garden work to world politics to latest UFO sightings. We approximate the experience of William Alcott, who wrote in 1837 of a meal he shared with friends: "With their plain dinner, and pleasant conversation, they pass half an hour or even more; sometimes sitting till 1 o'clock, especially if they have company. If the most illustrious visitors are present, they add nothing to the bread and potatoes, or whatever plain dishes they happen to have on the table, except perhaps some one kind of the best fruit of the season; and they never make any apologies" (*Ways of Living on Small Means*).

We have not yet mentioned that we do not eat meat. We regard vegetarianism as an essential part of the good life, in its ethical and humanitarian as well as health aspects. To us, leaving off flesh-eating is taken for granted. If not a requisite to living a good life, let's say it is to *us*. We cannot conceive of kindly, considerate, aware people consuming carcasses of perceptive, defenseless animals who were raised in captivity for slaughter. We think that the gross and selfish custom will be abandoned in years to come, as the eating of human flesh has been largely abandoned. If not for humanitarian or health reasons, then for economic and world-hunger reasons the vast population must eventually turn to vegetarianism. It is well known that the breeding and feeding of animals for slaughter uses more land than raising vegetable crops which can be eaten direct, without passing through animals' bodies.

For kindness, common sense, economy, simplicity, aesthetics, and health reasons, we eat nothing that walks or wiggles. This leaves a certain latitude as to eggs, milk, yogurt, cheese, honey, butter. We go very lightly on these animal products but do not eschew them altogether. We don't buy a dozen eggs a year, but eat them when served in food away from home. We haven't bought a quart of milk in years, though we get it in some foods served to us.

We eat yogurt and on festive occasions ice cream. Cheese, particularly cottage cheese, we may eat once or twice a week. We think vegetable oils better for us than butter, and they involve less exploitation of animals.

Taking honey from bees certainly exploits the bees, even if it does not starve or kill them; it robs them of the results of their intensive labor. We still eat honey but question it morally, as well as on account of its concentration of sweetness. On the whole, we go easy on sweetening, using a minimum of honey or maple syrup. We think most Americans overdo the sweet thing. They also overeat on protein and cereal starches. We live on a very low protein diet, with seeds, nuts and a few vegetables and fruits supplying what we get. Cereal starches we eat only once a day.

We prefer live foods and raw foods, eating them as fresh as possible, while they are still vibrant with life. If food is to be cooked, baking is better than boiling, and both better than frying. We try to keep the diet simple, with few mixtures, and preferably one food at a time. We have gone on mono-diets—for example, eating only apples for days on end, or subsisting on juices, or fasting on water only—for ten days at a time. We find this salutary, cleansing and restful for the body, let alone the easement for the housewife. One day a week we aim at twenty-four hours on just liquids, either juices or water. We enjoy these days of fasting and look forward to them as one of the high points of the week.

Not eating (so long as one is not starving) can be as enjoyable as eating. Just as shaving the head completely of hair can give one a godlike feeling of lack of clutter, so going without food can give one a feeling of freedom and release that is real emancipation.

Time and again we are queried on our sources of protein. Where, we ask in return, do the cows, elephants and rhinoceros—all sturdy, strong creatures—get their protein needs supplied? From grass, from foliage, from green things, is the answer. They do not eat refined, cooked, pickled, dehydrated, pasteurized, spiced, salted, canned, or otherwise conditioned materials. They are simple creatures eating nature's way. We try to be the same: simple-living people surviving on simple food, home grown, organically grown, and simply prepared. Our foods are vital, nutritional and economical. For fifty years we have thrived on such, and plan to continue so until our last days.

photo: Ralph T. Gardner

"There is no man nor woman the which have any respect to them selfe that can be a better Phesycion for theyr owne saveguarde than theyr owne selfe can be, to consyder what thynge the whiche doth them good, And to refrayne from such thynges that doth them hurte or harme."

Andrewe Boorde, A Dyetary of Helth, *1542*

"It is undeniably one of the most important Businesses of this Life to preserve our selves in Health. But there is scarce one Man or Woman of a thousand that does in earnest consider and pursue the means of preserving their Health, but either lives at Random, or at least takes up with the pernicious Notions of Custom, Tradition and Blind Guides, whose Prescriptions of Diet are most improper and prejudicial, their Medicines Nauseates to Nature, and their Physick a close Confederate with the invading Disease."

Thomas Tryon, The Good House-wife, *1692*

"It is most certain that 'tis easier to preserve Health than to recover it, and to prevent Diseases than to cure them."

Dr. George Cheyne, An Essay of Health and Long Life, *1725*

"We should strengthen and beautify and industriously mould our bodies to be fit companions of the soul, assist them to grow up like trees, and be agreeable and wholesome objects in nature."

Henry David Thoreau, Journal, *January 25, 1841*

"I have been asked sometimes how I could perform so large an amount of work with apparently so little diminution of strength. I attribute my power of endurance to a long-formed habit of observing, every day of my life, the simple laws of health, and none more than the laws of eating. It ceases any longer to be a matter of self-denial. It is almost like an instinct. I have made eating with regularity and with a reference to what I have to do, a habit so long that it ceases any longer to be a subject of thought. It almost takes care of itself. I attribute much of my ability to endure work to good habits of eating, constant attention to the laws of sleep, physical exercise, and cheerfulness."

Solon Robinson, Facts for Farmers, *1869*

"The whole of the material for growth comes from the food we eat. Wear and tear is made good from foodstuffs. There must be a fund to meet depreciation, and this depreciation is made good by food and food alone. The energy for work comes from the slow utilization and combustion of food in the cells of the body."

V. H. Mottram, Food and the Family, *1925*

"We are rarely ill, and if we are, we go off somewhere and eat grass until we feel better."

Paul Gallico, The Silent Miaow, *1964*

CHAPTER 17

WE PRACTICE HEALTH

WE LIVE in a society that takes sickness for granted and looks upon abounding health as lucky or exceptional. There are family doctors by the thousands all over the Western world who practice medicine by prescribing pills and drugs. Doctors (and dentists) are experts on sickness. Only a minority are practitioners of health.

Early in their history the Chinese are reported to have paid doctors their regular fees so long as people remained well. If and when a person became ill, the medical fees stopped until the patient was restored to health.

We, in the West, follow an opposite practice. We pay our doctors when we get sick and continue to pay them while we remain sick. The longer and the more serious the sickness, the longer the medical fees are paid. The lesson of such a situation is obvious to any reasoning being. If the doctor is paid when his clients are sick, the more often they are sick the more often the doctor will get a fee. Equally, the longer the sickness lasts and the more often the doctor is called in, the larger the income of the doctor.

Doctors often object to this reasoning. All the same, economic forces are potent factors in any society. Doctors, dentists and druggists and their families buy their groceries and pay their rent and taxes out of the money supplied by people who are in need of their services.

We cannot forget the story of the society lady who woke up one morning with a red blotch on her cheek. Much disturbed, she called up her family doctor and made an appointment to see him. The doctor looked carefully at the disfiguring spot and shook his head dubiously. "Is it that serious, doctor?" the lady questioned anxiously. "Rather serious, Madam," answered the doctor. "If you had waited for an hour or two before coming, the spot would have disappeared and I would have lost my fee." All doctors are not so frank.

The question "Who is your family doctor?" suggests that you expect to be sick and will therefore need help, so you keep a family doctor. The person who expects to be well does not need a doctor. He takes health for granted.

We homesteaded in Vermont for nineteen years without having a family doctor. We have homesteaded in Maine for more than twenty-five years equally free of permanent medical advice because we have been chronically well. In a word, we have practiced health by conducting ourselves in a way that keeps us well.

There is an old saying that practice makes perfect. Following this precept, those who want to be healthy must practice health. They must work at it. Like any other activity, the practice of health involves one or more choices. We choose to live quietly and simply, to exercise in the open air, to keep sensible hours and not overdo physically. We choose to exercise our bodies not in gymnasiums or on golf courses or tennis courts but doing useful outdoor physical work. We choose to live in the country rather than the city, with its polluted air, noise and stress. We prefer clean fresh air, sunshine, clear running water. We choose to cut our own fuel in our own woods rather than pay the oil barons. We design and construct our own buildings. We grow and prepare our own food, rather than shop in the supermarkets.

There are endless choices we all make daily and yearly—even hourly. One of our older neighbors came back home from an operation that was necessary to remove a cancer from his lip. Our friend was an inveterate chain cigarette smoker. He and the rest of his family had no doubt that cigarette smoking had resulted in the cancer. Back from his operation and surrounded by his family and friends, he reached for a cigarette, looked at it for a long moment, took out a match, snapped it with his thumbnail as was his wont and lit his first post-operational cigarette. He had made his choice. He preferred smoking to health. All choices are not equally as simple and single-tracked.

Early in life both of us faced up to a series of choices regarding health. These choices were never cut, dried and easy. Usually the decisions, especially about food and drink, forced us to separate ourselves from the customs of family, friends and acquaintances. We faced alternatives, in attitudes to life and current practices. Generally our young friends went along with the crowd: it was so easy to say yes, thank you. It took courage, knowledge and determination to say no, thank you, especially when some tasty edible was being offered. Early in our mid-teens we were faced with the choice to go along with the crowd or to choose habits that would establish and preserve health.

When we say that we practice health, or that we are healthy, we mean that the organism is operating according to plan and purpose as judged by its function and its product.

The human organism—physical, emotional, mental, spiritual—is the sum total of a human being alive and in good working order. This human organism has been evolving toward its present form on earth for at least two million years, and possibly a great deal longer. Over these vast reaches of time and sequence the processes of biological selection have matured the human organism into a wide range of variations of the present-day human being, with the same general contour, but differing in size, shape, color.

Human organisms are alive, growing and developing in successive stages of their life process or cycle. They generate their own energy patterns, converting energy intake into the wide variety of functions possessed by a complicated apparatus like the body, emotional structure, mind and spirit of a normal human being.

Like any other natural process, human energy is expressed in an aggregate of living cells, tissues, organs and members, which make up the human body. Not only are these coordinations highly complex, coordinated, specialized fragments of a complex whole, but, so far as we are able to determine, no two of these specialized organisms are identical in form, structure or function. Each is unique, living its own individual life span under its own individual direction—similar, but never identical.

The life processes we see operating in humanity at the present moment have been repeated numberless times in the past and are being lived (repeated) in the present, similar but not uniform in manner, each integer or life unit possessing its own unique identity through the minutes, hours, days, years, centuries and millennia.

It would be a mistake to lay too much stress on the multiplicity of individual life acts, stages and processes. On the other hand, it is impossible to pass the life process by without calling attention, with great admiration, to its multiplicity and its diversity and its continuity.

We make these observations in a country that abounds in complex mechanisms, each with its own multitudes of standard replacement parts and fittings—sewing machines, generators, typewriters, planting and harvesting machines. Doubly impressive is an organism that is as complicated, that is alive and capable of reproducing itself endlessly, given the necessary energy and raw materials. The result is trebly impressive when we note that the human organism cannot only duplicate itself arithmetically, but that it can proceed analytically and critically evaluating the product during the operation.

When a mechanism can turn out a needed or wanted product, we say it is in good working order. When an organism can do this, we say it is in good health. That is, the organism is producing the results that it was developed to produce. Through direct and indirect experience we convert a more or less helpless infant at birth into a trained, skilled, experienced adult capable of performing one or more of the multitude of operations necessary to produce a work of art, a telephone system or a suspension bridge.

The important factor in the situation is that we are consciously aware of the entire process, know when and why we fail and are increasingly certain that if we follow the necessary procedures we can establish and maintain the quality and length of our lives. In other words, we can avoid sickness and ill health by adopting life styles that bypass illness and stabilize good health.

We are proclaiming human health as the condition of body, emotions, mind and spirit that all self-respecting human beings should strive to achieve. We believe that we should practice or maintain health as we practice any other human virtue. On second thought, health should probably head the list of primary objectives toward which humans should strive.

We are not professional healthists. We are mere laymen. Our claim to champion the cause of health is based on four noteworthy facts: first, both of us have been well and have practiced health since we were in our teens. Second, both of us are now healthy in old age. Third, we have and have had no family doctor. Fourth, we are active in organizations that advocate and practice health.

We round out this chapter by setting down some of the essential aspects in the practice of health.

1. Sufficient knowledge and experience are necessary to know what practice of health involves. Scott's mother was a health fan at the end of the nineteenth century, with six children to bring up. As high-school students, Scott's generation was exposed to the teachings of Bernarr MacFadden and other health enthusiasts of the period, who published literature, gave lectures and held classes and workshops in which they called into question the health pattern of the Establishment on the ground that it accepted sickness as an essential feature of human life and maintained an elaborate apparatus of institutions and trained personnel which was paid and financed for looking after sick people. Helen's parents were theosophists, involved in Eastern philosophies, were vegetarians and believed in healthful practices.

2. Instead of the negative attitude of treating sickness after the organism breaks down, the healthist proposes practices that make people well and keep them well. Such a view makes illness subnormal and to be replaced by practices that establish and maintain health as normal.

3. Health therefore should be taught and practiced consistently as an essential feature of good citizenship. This means, of course, that doctors and other health professionals should set an example to their patients by being well and staying well. Health workers should be living and shining examples of abounding health: Physician, heal thyself.

4. Since the human being is fueled or energized by taking nourishment into the alimentary canal, healthists should know what solids and liquids are necessary to maintain health and see to it that the food intake corresponds to the best that health science has attained. Food intake should be properly balanced chemically. It should be fresh, free of poisons or deleterious substances. Food should be whole and unprocessed where possible (cooking is a form of processing). Food should be supplied in proper quantity as well as be of proper quality. Since

the body is built up and its cells are replaced by the food-drink intake, the greatest care should be taken by health authorities to see that only the best of building materials are supplied to the citizenry—especially to the children of today who are the adults of tomorrow.

5. Health depends not alone upon food and drink. Sunshine, fresh air, pure water and exercise are equally important elements in health. Balanced, healthful living requires at-oneness with all aspects of nature.

6. Sufficient shelter, good housing and adequate clothing are essential to health, especially in climatic belts where bad weather is encountered.

7. Tension is to be avoided, in the interest of health. This is particularly the case in urban living under conditions of congestion, high speed, noise and other irritants which impair and undermine human well-being. Those who advocate health should live wisely and sanely, socially as well as bodily.

photo: Louis Penfield

"O happy who thus liveth! Not caring much for gold;
With clothing which sufficeth to keep him from the cold.
Though poor and plain his diet, yet merry 'tis, and quiet."

Elizabethan Song Book, *circa 1588*

"I never had any other desire so strong, as so like to covetousness as that one, which I have had always, that I might bee master at last, of a small House and large Garden, with very moderate conveniences joyned to them, and there dedicate the remainder of my life, onely to the culture of them, and study of Nature. And there, with no design beyond my wall, whole and entire to lye, in no unactive ease, and no unglorious poverty."

Abraham Cowley, The Garden, *1666*

"It must be owned, indeed, that the Town hath its Pleasures as well as the Countrey. But how alluring soever the Pleasures of the Town may seem to us, whilst Health and Strength and the Gaities of Youth last, Envy, Malice and Double-dealing do so frequent the most busy Parts of the World, which tend to mar all those Delights. We shall be inclined to declare in Favour of the innocent Simplicity of a Countrey-life. As long as the World lasts, the Pleasures and Entertainments which Gardening and Agriculture afford, will be the pursuit of wise Men."

John Laurence, A New System of Agriculture, *1726*

"Our life is a busy round of a great variety of occupations, all tending to health and chearfulness. We rise every day with the sun, and in the cool of morning, employ ourselves in business which requires some strength. The garden takes up much of our time. In the afternoon we read and work. In the evening I take to my tools and labour again, either hoeing, digging, chopping wood against winter, or any work of the season that is necessary. Such generally is the round of the day.... Our happy little farm yields us a constant amusement of a most rational and agreeable kind."

Arthur Young, The Adventures of Emmera, *1767*

"Countrymen in general are a very happy people; they enjoy many of the necessities of life upon their own farms, and what they do not so gain, they have from the sale of their surplus products; it is remarkable to see such numbers of these men in a state of great ease and content, possessing all the necessaries of life, but few of the luxuries of it."

Anonymous, American Husbandry, *1775*

Chapter 18

A REWARDING WAY TO LIVE

LIKE multitudes of people all over the world, we are seeking a good life—a simple, balanced, satisfying life style. Like them, our aim is to lend a hand in shaping the planet into a homelike living place for successive generations of human beings and for the many other life forms domiciled in and on Mother Earth, her lands and waters.

Immediate needs for a good life are food and shelter, as a basis for survival. Beyond these basic necessaries are amenities like education, recreation and travel, which make life more satisfying and rewarding for individuals and small local groups such as families and other collectives.

We begin our listing of good life attributes with our four-four-four formula: four hours of bread labor; four hours of professional activity; and four hours dedicated to fulfilling our obligations and responsibilities as members of the human race and as participants in various local, regional, national and world civic activities.

Bread labor provides the basic essentials of living normal, healthful, serviceful lives. The work of the world must be done and we should all share in it. Professional activities enable us to specialize and contribute our mite to the world's sum total of skills and competencies. Association enables us to share experience and knowledge with our fellow beings.

The four-four-four formula should be specific as well as general. Everyone, rich or poor, young or old, can contribute somewhat to the world's physical work. Bread labor can and should be performed by every able-bodied human being from age 7 to 77 (though Scott at 95 is still carrying his end of the load). Bread labor should be an obligatory and honorable phase of the daily routine in which everyone can take an active part as a matter of course. This daily contribution to the work of the world will make a vigorous, self-supporting society.

The personal interests and skills of each human being will be another contribution that will produce inventions and artistic achievements from which society will also benefit.

Civic responsibilities and acitivites taken on by all adults will benefit the whole society and bring people together in common interests of survival and well-being.

If we would have things done, we must be prepared to do them. This principle applies to the life of the individual; it is even more urgent when issues of group concern are up for consideration. Those who would be well served serve themselves, individually and collectively.

With us Westerners in the present century this principle deserves double emphasis, because in our immediate past we have overemphasized the individual to the detriment of the group. Groups have grown larger, more numerous and more complex. Doubling populations automatically shift attention from individual to group activity. Increase in size, coupled with growing complexity and interdependence, must play down the individual and upgrade the group and group problems.

Since the Mexican Revolution of 1910, through the Russian Revolution of 1917, the East European, Asian and African revolutions that followed the wars of 1914, 1936 and 1945, and the upheaval in China that began as early as 1899, a third of mankind has launched a movement for collectivist socialist-communist action that places group welfare above individual interests. Up to this point of social evolution, life, liberty and the pursuit of happiness in terms of eighteenth-century perspective dominated the planet-wide struggle for a better life. The socialist-communist revolutions of the present century subordinate individual happiness for the bourgeoisie to well-being for the entire group. In the face of growing sociological challenges, we in the West have continued to underscore the eighteenth-century formula of freedom, independence, and self-determination.

Through the years, the authors of this book have seen steady and persistent efforts by individuals, and by social groups, to protect and conserve the natural environment. There are vast regions in the socialist countries and in Scandinavia and other parts of Europe in which conservation of nature is a first charge. This is an aspect of civic responsibility that should come to the fore more generally.

The essence of life consists in living. In the words of Robert Louis Stevenson, "To travel hopefully is better than to arrive, and the true success is to labor."

From earliest childhood to the final insecure steps of old age, those who put the most into life get the most out of life. This applies to quantity of life and quantity of output. Theory guides; practice determines. The uniting of theory with practice provides a higher degree of assurance and promotes a more rewarding body of dependable guidance for individual and group living.

Muscles grow strong and responsive with exercise. Muscles of the spectator go flabby and shrink. This rule is equally applicable to the problems of physical function and social action.

Personally, we in our entire homesteading venture have endeavored to keep our social as well as physical muscles in shape. We tried, as a couple, and insofar as we could in groups, to set up and continue a life pattern to maintain health and sanity in a period of social insecurity, conflict, disruption and disintegration.

We began experimenting with an alternative life pattern nearly half a century ago, in 1932. We were not young, but we were adventurous. Our first steps were tentative. As we proceeded, we became clearer in our thinking and surer that the course we were following was right for us.

We were trying out a life style that was not new in history, but was new in our generation. We left city living, with its civilized polish and its murky poverty, and launched out into a simpler, more self-sufficient life in the country.

Our general aim was to set up a use economy for ourselves independent of the established market economy and for the most part under our own control, thereby freeing ourselves from undue dependence on the Establishment.

We wanted to provide ourselves with the economic means that would free at least a third of our time and energy to carry on our professional work and our interest in improving the social environment.

Specifically, we have provided ourselves during more than four decades with the basic necessaries of life in exchange for a sufficient amount of planning, persistence and hard work.

We have been able to carry on our writing and research. Scott has written six books since leaving the city, only one of which saw the light of day in the commercial book market. Six additional books were written jointly with Helen. She has continued her lifelong interest in music and has added secretarial, editorial, writing and house-building skills to her accomplishments.

In our forty-odd years of homesteading we have reshaped old ideas and practices and tried out new ones. Never for a moment did we sit back and say to ourselves: This is it; we have learned all there is to know about homesteading; we have arrived.

On the contrary, the further we went with our bread labor, our professional activities and our civic work, the more we realized that we had only scratched life's surface. There is so much more to experience, to do and to learn that only many lives ahead will give us time and opportunities to fulfill all we hope to do.

We have done our best to contribute to the knowledge and possibilities of homesteading in New England for four decades.

We have been able, from our home base, to initiate a considerable number of young people into the possibilities of learning a way of self-sufficient living. We helped them to get started on a life style more satisfying than that of the average young American.

Through writings, speaking, workshops and other means of reaching the public, we have tried to show what can be accomplished by oldsters.

We have manifested that health and vigor can be maintained to a degree consistent with efforts to live simply and sanely.

We have done what we could to conserve and improve the natural environment and to make its facilities available for social advancement.

Driven by whatever urge, motivated by a wide variety of interests and convictions, all human beings, early in their lives, are possessed by the wish and the will to live a satisfying and rewarding life. Speaking for ourselves, we, at advanced ages, are still questioning, investigating, searching and aiming to build a more rewarding and more creative life. For those who are so minded and so willed, we have written this book.

If we have helped a number of people to get started on a life style more satisfying than that of the average United States citizen, they have in turn contributed measurably to the building of our own lives. They have helped us with gardening and with building. By the exchange of insights, experiences and skills, they have cooperated as fellow workers in our joint efforts. We thank them and salute them as we continue in our own varied fields of productive and creative endeavor.

EPILOGUE

"Truly I cannot but here take Occasion to exhort all Philosophic Gentlemen to employ a reasonable Share of their Thoughts and Experiments on the Subject of Agriculture as a more becoming Exercise and Relaxation than Hunting or Cards; and to be sure, more conducing to the Health of the Body, the Strength of the Mind, and to the Capacity of Generosity in the Fortune, than many other fashionable but criminal Excesses. For it ought to be observed that it is an Employment which will at once contract their Wants, and give a larger Ability to supply them; 'twill give greater Relish to the Enjoyments of Life, and make every Part thereof sweetly varied between Ease and delightful Labour."

John Laurence, A New System of Agriculture, *1726*

"The reader must not take it for granted that in going into the country we escaped all the annoyances of domestic life peculiar to the city, or that we fell heir to no new ones, such as we had never before experienced. He must remember that this is a world of compensations, and that nowhere will he be likely to find either an unmixed good or an unmixed bad. Such was exactly our experience. But on summing up the two, the balance was decidedly in our favor."

Anonymous, Ten Acres Enough, *1864*

"This book is the fruit of years of labor in a great and good field. It certainly contains much that will be useful to all classes who till the earth or live in farmers' houses....Though not perfect, farmers (and their wives) will find this book a useful one. If not invaluable, I hope it is one that they cannot afford to do without....Usefulness instead of elegance has been aimed at....To those who know the name of the author (and the number is large) I hope this book will be a welcome bequest. I hope it will be the means through which that name may live in love and honor with your children and children's children around many an American hearthstone."

Solon Robinson, Facts for Farmers, *1869*

"If we had ample means and could choose any kind of life we wished, we would choose what we have chosen. And when I say we, I mean we. There are many differences between a man's viewpoint and a woman's, even though they may live side by side in the same house year in and year out. But there must be a profound unshaken unity underneath the difference if they are to make a success of such a life as we have lived, because the things that must be passed by are things that one or the other might consider indispensable. As for children, I cannot help but think that they gain far more than they lose, in happiness and experience. By and large, it is the best life for children. And later, they must make their own choice."

Gove Hambidge, Enchanted Acre: Adventures in Backyard Farming, *1935*

INDEX